Study Guide and Intervention Workbook

Algebra 2

$f(x) = -0.5x^2$

McGraw Hill Glencoe

To the Student

This *Study Guide and Intervention Workbook* gives you additional examples and problems for the concept exercises in each lesson. The exercises are designed to aid your study of mathematics by reinforcing important mathematical skills needed to succeed in the everyday world. The materials are organized by chapter and lesson, with two Study Guide and Intervention worksheets for every lesson in *Glencoe Algebra 2*.

Always keep your workbook handy. Along with your textbook, daily homework, and class notes, the completed *Study Guide and Intervention Workbook* can help you in reviewing for quizzes and tests.

To the Teacher

These worksheets are the same ones found in the Chapter Resource Masters for *Glencoe Algebra 2*. The answers to these worksheets are available at the end of each Chapter Resource Masters booklet as well as in your Teacher Wraparound Edition interleaf pages.

The McGraw·Hill Companies

 Glencoe

Send all inquiries to:
Glencoe/McGraw-Hill
8787 Orion Place
Columbus, OH 43240-4027

ISBN: 978-0-07-890861-3
MHID: 0-07-890861-2

Printed in the United States of America.

12 13 14 15 16 17 18 19 RHR 20 19 18 17 16

Study Guide and Intervention Workbook, Algebra 2

Contents

NAME _____ DATE _____ PERIOD _____

1-1 Study Guide and Intervention

Expressions and Formulas

Order of Operations

Order of Operations	Step 1	Evaluate expressions inside grouping symbols.
	Step 2	Evaluate all powers.
	Step 3	Multiply and/or divide from left to right.
	Step 4	Add and/or subtract from left to right.

Example 1 Evaluate $[18 - (6 + 4)] \div 2$.

$$[18 - (6 + 4)] \div 2 = [18 - 10] \div 2$$
$$= 8 \div 2$$
$$= 4$$

Example 2 Evaluate $3x^2 + x(y - 5)$ if $x = 3$ and $y = 0.5$.

Replace each variable with the given value.

$$3x^2 + x(y - 5) = 3 \cdot (3)^2 + 3(0.5 - 5)$$
$$= 3 \cdot (9) + 3(-4.5)$$
$$= 27 - 13.5$$
$$= 13.5$$

Exercises

Evaluate each expression.

1. $14 + (6 \div 2)$

2. $11 - (3 + 2)^2$

3. $2 + (4 - 2)^3 - 6$

4. $9(3^2 + 6)$

5. $(5 + 2^3)^2 - 5^2$

6. $5^2 + \frac{1}{4} + 18 \div 2$

7. $\dfrac{16 + 2^3 \div 4}{1 - 2^2}$

8. $(7 - 3^2)^2 + 6^2$

9. $20 \div 2^2 + 6$

10. $12 + 6 \div 3 - 2(4)$

11. $14 \div (8 - 20 \div 2)$

12. $6(7) + 4 \div 4 - 5$

13. $8(4^2 \div 8 - 32)$

14. $\dfrac{6 + 4 \div 2}{4 \div 6 - 1}$

15. $\dfrac{6 + 9 \div 3 + 15}{8 - 2}$

Evaluate each expression if $a = 8.2$, $b = -3$, $c = 4$, and $d = -\frac{1}{2}$.

16. $\dfrac{ab}{d}$

17. $5(6c - 8b + 10d)$

18. $\dfrac{c^2 - 1}{b - d}$

19. $ac - bd$

20. $(b - c)^2 + 4a$

21. $\dfrac{a}{d} + 6b - 5c$

22. $3\left(\dfrac{c}{d}\right) - b$

23. $cd + \dfrac{b}{d}$

24. $d(a + c)$

25. $a + b \div c$

26. $b - c + 4 \div d$

27. $\dfrac{a}{b + c} - d$

1

1-1   Study Guide and Intervention   *(continued)*

Expressions and Formulas

Formulas A **formula** is a mathematical sentence that expresses the relationship between certain quantities. If you know the value of every variable in the formula except one, you can use substitution and the order of operations to find the value of the remaining variable.

Example   The formula for the number of reams of paper needed to print n copies of a booklet that is p pages long is $r = \dfrac{np}{500}$, where r is the number of reams needed. How many reams of paper must you buy to print 172 copies of a 25-page booklet?

$r = \dfrac{np}{500}$ Formula for paper needed

$= \dfrac{(172)(25)}{500}$ $n = 172$ and $p = 25$

$= \dfrac{4300}{500}$ Evaluate (172)(25).

$= 8.6$ Divide.

You cannot buy 8.6 reams of paper. You will need to buy 9 reams to print 172 copies.

Exercises

1. For a science experiment, Sarah counts the number of breaths needed for her to blow up a beach ball. She will then find the volume of the beach ball in cubic centimeters and divide by the number of breaths to find the average volume of air per breath.

   a. Her beach ball has a radius of 9 inches. First she converts the radius to centimeters using the formula $C = 2.54I$, where C is a length in centimeters and I is the same length in inches. How many centimeters are there in 9 inches?

   b. The volume of a sphere is given by the formula $V = \dfrac{4}{3}\pi r^3$, where V is the volume of the sphere and r is its radius. What is the volume of the beach ball in cubic centimeters? (Use 3.14 for π.)

   c. Sarah takes 40 breaths to blow up the beach ball. What is the average volume of air per breath?

2. A person's basal metabolic rate (or BMR) is the number of calories needed to support his or her bodily functions for one day. The BMR of an 80-year-old man is given by the formula $\text{BMR} = 12w - (0.02)(6)12w$, where w is the man's weight in pounds. What is the BMR of an 80-year-old man who weighs 170 pounds?

1-2 Study Guide and Intervention

Properties of Real Numbers

Real Numbers All real numbers can be classified as either rational or irrational. The set of rational numbers includes several subsets: natural numbers, whole numbers, and integers.

R	real numbers	{all rationals and irrationals}
Q	rational numbers	{all numbers that can be represented in the form $\frac{m}{n}$, where m and n are integers and n is not equal to 0}
I	irrational numbers	{all nonterminating, nonrepeating decimals}
Z	integers	{..., −3, −2, −1, 0, 1, 2, 3, ...}
W	whole numbers	{0, 1, 2, 3, 4, 5, 6, 7, 8, ...}
N	natural numbers	{1, 2, 3, 4, 5, 6, 7, 8, 9, ...}

Example Name the sets of numbers to which each number belongs.

a. $-\frac{11}{3}$ rationals (Q), reals (R)

b. $\sqrt{25}$

$\sqrt{25} = 5$ naturals (N), wholes (W), integers (Z), rationals (Q), reals (R)

Exercises

Name the sets of numbers to which each number belongs.

1. $\frac{6}{7}$ **2.** $-\sqrt{81}$ **3.** 0   **4.** 192.0005

5. 73   **6.** $34\frac{1}{2}$ **7.** $\frac{\sqrt{36}}{9}$ **8.** 26.1

9. π **10.** $\frac{15}{3}$ **11.** $-4.\overline{17}$

12. $\frac{\sqrt{25}}{2}$ **13.** −1   **14.** $\sqrt{42}$

15. −11.2   **16.** $-\frac{8}{13}$ **17.** $\frac{\sqrt{5}}{2}$

18. $33.\overline{3}$ **19.** 894,000   **20.** −0.02

1-2 Study Guide and Intervention (continued)

Properties of Real Numbers

Properties of Real Numbers

Real Number Properties		
For any real numbers a, b, and c		
Property	**Addition**	**Multiplication**
Commutative	$a + b = b + a$	$a \cdot b = b \cdot a$
Associative	$(a + b) + c = a + (b + c)$	$(a \cdot b) \cdot c = a \cdot (b \cdot c)$
Identity	$a + 0 = a = 0 + a$	$a \cdot 1 = a = 1 \cdot a$
Inverse	$a + (-a) = 0 = (-a) + a$	$a \cdot \frac{1}{a} = 1 = \frac{1}{a} \cdot a, a \neq 0.$
Closure	$a + b$ is a real number.	$a \cdot b$ is a real number.
Distributive	$a(b + c) = ab + ac$ and $(b + c)a = ba + ca$	

Example Simplify $9x + 3y + 12y - 0.9x.$

$\begin{aligned} 9x + 3y + 12y - 0.9x &= 9x + (-0.9x) + 3y + 12y && \text{Commutative Property (+)} \\ &= (9 + (-0.9))x + (3 + 12)y && \text{Distributive Property} \\ &= 8.1x + 15y && \text{Simplify.} \end{aligned}$

Exercises

Simplify each expression.

1. $8(3a - b) + 4(2b - a)$

2. $40r + 18t - 5t + 11r$

3. $\frac{1}{5}(4j + 2k - 6j + 3k)$

4. $10(6g + 3h) + 4(5g - h)$

5. $12\left(\frac{a}{3} - \frac{b}{4}\right)$

6. $8(2.4r - 3.1t) - 6(1.5r + 2.4t)$

7. $4(20 - 4p) - \frac{3}{4}(4 - 16p)$

8. $5.5j + 8.9k - 4.7k - 10.9j$

9. $1.2(7x - 5y) - (10y - 4.3x)$

10. $9(7d - 4f) - 0.6(d + 5f)$

11. $2.5(12m - 8.5p)$

12. $\frac{3}{4}p - \frac{1}{5}r - \frac{3}{5}r - \frac{1}{2}p$

13. $4(10g + 80h) - 20(10h - 5g)$

14. $2(15d + 45c) + \frac{5}{6}(12d + 18c)$

15. $(7y - 2.1x)3 + 2(3.5x - 6y)$

16. $\frac{2}{3}(18m - 6p + 12m + 3p)$

17. $14(j - 2k) - 3j(4 - 7k)$

18. $50(3a - b) - 20(b - 2a)$

1-3 Study Guide and Intervention
Solving Equations

Verbal Expressions and Algebraic Expressions

The chart suggests some ways to help you translate word expressions into algebraic expressions. Any letter can be used to represent a number that is not known.

Word Expression	Operation
and, plus, sum, increased by, more than	addition
minus, difference, decreased by, less than	subtraction
times, product, of $\left(\text{as in } \frac{1}{2} \text{ of a number}\right)$	multiplication
divided by, quotient	division

Example 1 Write an algebraic expression to represent 18 less than the quotient of a number and 3.

$\dfrac{n}{3} - 18$

Example 2 Write a verbal sentence to represent $6(n - 2) = 14$.

Six times the difference of a number and two is equal to 14.

Exercises

Write an algebraic expression to represent each verbal expression.

1. the sum of six times a number and 25

2. four times the sum of a number and 3

3. 7 less than fifteen times a number

4. the difference of nine times a number and the quotient of 6 and the same number

5. the sum of 100 and four times a number

6. the product of 3 and the sum of 11 and a number

7. four times the square of a number increased by five times the same number

8. 23 more than the product of 7 and a number

Write a verbal sentence to represent each equation.

9. $3n - 35 = 79$

10. $2(n^3 + 3n^2) = 4n$

11. $\dfrac{5n}{n + 3} = n - 8$

1-3 Study Guide and Intervention *(continued)*

Solving Equations

Properties of Equality To solve equations, we can use properties of equality.

Addition and Subtraction Properties of Equality	For any real numbers a, b, and c, if $a = b$, then $a + c = b + c$ and $a - c = b - c$.
Multiplication and Division Properties of Equality	For any real numbers a, b, and c, if $a = b$, then $a \cdot c = b \cdot c$ and, if $c \neq 0$, $\frac{a}{c} = \frac{b}{c}$.

Example 1 Solve $10 - 8x = 50$.

$10 - 8x = 50$	Original equation
$10 - 8x - 10 = 50 - 10$	Subtract 10 from both sides.
$-8x = 40$	Simplify.
$x = -5$	Divide both sides by -8.

Example 2 Solve $4x + 5y = 100$ for y.

$4x + 5y = 100$	Original equation
$4x + 5y - 4x = 100 - 4x$	Subtract $4x$ from both sides.
$5y = 100 - 4x$	Simplify.
$y = \frac{1}{5}(100 - 4x)$	Divide both sides by 5.
$y = 20 - \frac{4}{5}x$	Apply the distributive property.

Exercises

Solve each equation. Check your solution.

1. $3s = 45$

2. $17 = 9 - a$

3. $5t - 1 = 6t - 5$

4. $\frac{2}{3}m = \frac{1}{2}$

5. $7 - \frac{1}{2}x = 3$

6. $-8 = -2(z + 7)$

7. $0.2b = 10$

8. $3x + 17 = 5x - 13$

9. $5(4 - k) = -10k$

10. $120 - \frac{3}{4}y = 60$

11. $\frac{5}{2}n = 98 - n$

12. $4.5 + 2p = 8.7$

13. $4n + 20 = 53 - 2n$

14. $100 = 20 - 5r$

15. $2x + 75 = 102 - x$

Solve each equation or formula for the specified variable.

16. $a = 3b - c$, for b

17. $\frac{s}{2t} = 10$, for t

18. $h = 12g - 1$, for g

19. $\frac{3pq}{r} = 12$, for p

20. $2xy = x + 7$, for x

21. $\frac{d}{2} + \frac{f}{4} = 6$, for f

22. $3(2j - k) = 108$, for j

23. $3.5s - 42 = 14t$, for s

24. $\frac{m}{n} + 5m = 20$, for m

25. $4x - 3y = 10$, for y

1-4 Study Guide and Intervention

Solving Absolute Value Equations

Absolute Value Expressions The **absolute value** of a number is its distance from 0 on a number line. The symbol $|x|$ is used to represent the absolute value of a number x.

Absolute Value	• **Words**	For any real number a, if a is positive or zero, the absolute value of a is a. If a is negative, the absolute value of a is the opposite of a.
	• **Symbols**	For any real number a, $|a| = a$, if $a \geq 0$, and $|a| = -a$, if $a < 0$.

Example 1 Evaluate $|-4| - |-2x|$ if $x = 6$.

$$
\begin{aligned}
|-4| - |-2x| &= |-4| - |-2 \cdot 6| \\
&= |-4| - |-12| \\
&= 4 - 12 \\
&= -8
\end{aligned}
$$

Example 2 Evaluate $|2x - 3y|$ if $x = -4$ and $y = 3$.

$$
\begin{aligned}
|2x - 3y| &= |2(-4) - 3(3)| \\
&= |-8 - 9| \\
&= |-17| \\
&= 17
\end{aligned}
$$

Exercises

Evaluate each expression if $w = -4$, $x = 2$, $y = \frac{1}{2}$, and $z = -6$.

1. $|2x - 8|$

2. $|6 + z| - |-7|$

3. $5 + |w + z|$

4. $|x + 5| - |2w|$

5. $|x| - |y| - |z|$

6. $|7 - x| + |3x|$

7. $|w - 4x|$

8. $|wz| - |xy|$

9. $|z| - 3|5yz|$

10. $5|w| + 2|z - 2y|$

11. $|z| - 4|2z + y|$

12. $10 - |xw|$

13. $|6y + z| + |yz|$

14. $3|wx| + \frac{1}{4}|4x + 8y|$

15. $7|yz| - 30$

16. $14 - 2|w - xy|$

17. $|2x - y| + 5y$

18. $|xyz| + |wxz|$

19. $z|z| + x|x|$

20. $12 - |10x - 10y|$

21. $\frac{1}{2}|5z + 8w|$

22. $|yz - 4w| - w$

23. $\frac{3}{4}|wz| + \frac{1}{2}|8y|$

24. $xz - |xz|$

1-4 Study Guide and Intervention *(continued)*

Solving Absolute Value Equations

Absolute Value Equations Use the definition of absolute value to solve equations containing absolute value expressions.

> For any real numbers a and b, where $b \geq 0$, if $|a| = b$ then $a = b$ or $a = -b$.

Always check your answers by substituting them into the original equation. Sometimes computed solutions are not actual solutions.

Example Solve $|2x - 3| = 17$. Check your solutions.

Case 1	$a = b$	Case 2	$a = -b$
	$2x - 3 = 17$		$2x - 3 = -17$
	$2x - 3 + 3 = 17 + 3$		$2x - 3 + 3 = -17 + 3$
	$2x = 20$		$2x = -14$
	$x = 10$		$x = -7$

CHECK $|2x - 3| = 17$

$|2(10) - 3| \stackrel{?}{=} 17$

$|20 - 3| \stackrel{?}{=} 17$

$|17| \stackrel{?}{=} 17$

$17 = 17$ ✓

CHECK $|2x - 3| = 17$

$|2(-7) - 3| \stackrel{?}{=} 17$

$|-14 - 3| \stackrel{?}{=} 17$

$|-17| \stackrel{?}{=} 17$

$17 = 17$ ✓

There are two solutions, 10 and −7.

Exercises

Solve each equation. Check your solutions.

1. $|x + 15| = 37$

2. $|t - 4| - 5 = 0$

3. $|x - 5| = 45$

4. $|m + 3| = 12 - 2m$

5. $|5b + 9| + 16 = 2$

6. $|15 - 2k| = 45$

7. $5n + 24 = |8 - 3n|$

8. $|8 + 5a| = 14 - a$

9. $\frac{1}{3}|4p - 11| = p + 4$

10. $|3x - 1| = 2x + 11$

11. $\left|\frac{1}{3}x + 3\right| = -1$

12. $40 - 4x = 2|3x - 10|$

13. $5f - |3f + 4| = 20$

14. $|4b + 3| = 15 - 2b$

15. $\frac{1}{2}|6 - 2x| = 3x + 1$

16. $|16 - 3x| = 4x - 12$

1-5 Study Guide and Intervention

Solving Inequalities

One-Step Inequalities The following properties can be used to solve inequalities.

Addition and Subtraction Properties for Inequalities	Multiplication and Division Properties for Inequalities
For any real numbers a, b, and c: If $a < b$, then $a + c < b + c$ and $a - c < b - c$. If $a > b$, then $a + c > b + c$ and $a - c > b - c$.	For any real numbers a, b, and c, with $c \neq 0$: If c is positive and $a < b$, then $ac < bc$ and $\frac{a}{c} < \frac{b}{c}$. If c is positive and $a > b$, then $ac > bc$ and $\frac{a}{c} > \frac{b}{c}$. If c is negative and $a < b$, then $ac > bc$ and $\frac{a}{c} > \frac{b}{c}$. If c is negative and $a > b$, then $ac < bc$ and $\frac{a}{c} < \frac{b}{c}$.

These properties are also true for $\leq$ and $\geq$.

Example 1 Solve $2x + 4 > 36$. Graph the solution set on a number line.

$$2x + 4 - 4 > 36 - 4$$
$$2x > 32$$
$$x > 16$$

The solution set is $\{x \mid x > 16\}$.

13 14 15 16 17 18 19 20 21

Example 2 Solve $17 - 3w \geq 35$. Graph the solution set on a number line.

$$17 - 3w \geq 35$$
$$17 - 3w - 17 \geq 35 - 17$$
$$-3w \geq 18$$
$$w \leq -6$$

The solution set is $\{w \mid w \leq -6\}$.

−9 −8 −7 −6 −5 −4 −3 −2 −1

Exercises

Solve each inequality. Then graph the solution set on a number line.

1. $7(7a - 9) \leq 84$

2. $3(9z + 4) > 35z - 4$

3. $5(12 - 3n) < 165$

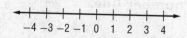

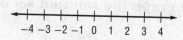

−8 −7 −6 −5 −4 −3 −2 −1 0

4. $18 - 4k < 2(k + 21)$

5. $4(b - 7) + 6 < 22$

6. $2 + 3(m + 5) \geq 4(m + 3)$

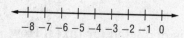

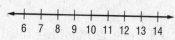

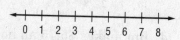

7. $4x - 2 > -7(4x - 2)$

8. $\frac{1}{3}(2y - 3) > y + 2$

9. $2.5d + 15 \leq 75$

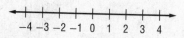

−14   −12   −10   −8   −6

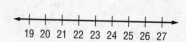

1-5 Study Guide and Intervention *(continued)*

Solving Inequalities

Multi-Step Inequalities An inequality is a statement that involves placing the inequality sign between two expressions. In order to solve the inequality, you need to find the set of all the values of the variable that makes the inequality true.

Example GAMES **After three quarters of the season has past, the Tigers have won 48 out of 72 games. How many of the remaining games must they win in order to win more than 70% of all their games this season?**

Understand   Let x be the number of remaining games that the Tigers must win. The total number of games they will have won by the end of the season is $\frac{3}{4}(48 + x)$. They should win at least 70% of their games.

Plan $\frac{3}{4}(48 + x) > 0.7(72)$

Solve $\frac{3}{4}(48 + x) > 0.7(72)$ Original Inequality

           $48 + x > \frac{4}{3}0.7(72)$ Multiply each side by $\frac{4}{3}$.

           $48 + x > 67.2$ Simplify.

           $x > 19.2$ Subtract 48 from each side.

The Tigers have to win 20 or more games.

Exercises

Solve each inequality. Then graph the solution set on a number line.

1. $c \geq \dfrac{c + 4}{3}$ **2.** $r + 7 < 3(2r - 6)$ **3.** $3h < \dfrac{2h + 26}{5}$

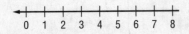   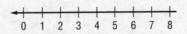   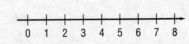

4. Jim makes $5.75 an hour. Each week, 26% of his total pay is deducted for taxes. How many hours does Jim have to work if he wants his take-home pay to be at least $110 per week? Write and solve an inequality for this situation.

1-6 Study Guide and Intervention

Solving Compound and Absolute Value Inequalities

Compound Inequalities A compound inequality consists of two inequalities joined by the word *and* or the word *or*. To solve a compound inequality, you must solve each part separately.

And Compound Inequalities	The graph is the intersection of solution sets of two inequalities.	Example: $x > -4$ and $x < 3$![number line](−5 −4 −3 −2 −1 0 1 2 3 4 5)
Or Compound Inequalities	The graph is the union of solution sets of two inequalities.	Example: $x \le -3$ or $x > 1$![number line](−5 −4 −3 −2 −1 0 1 2 3 4 5)

Example 1 Solve $-3 \le 2x + 5 \le 19$. Graph the solution set on a number line.

$$-3 \le 2x + 5 \quad \text{and} \quad 2x + 5 \le 19$$
$$-8 \le 2x \qquad\qquad\quad 2x \le 14$$
$$-4 \le x \qquad\qquad\quad\quad x \le 7$$
$$-4 \le x \le 7$$

−8 −6 −4 −2 0 2 4 6 8

Example 2 Solve $3y - 2 \ge 7$ or $2y - 1 \le -9$. Graph the solution set on a number line.

$$3y - 2 \ge 7 \quad \text{or} \quad 2y - 1 \le -9$$
$$3y \ge 9 \quad \text{or} \qquad 2y \le -8$$
$$y \ge 3 \quad \text{or} \qquad\quad y \le -4$$

−8 −6 −4 −2 0 2 4 6 8

Exercises

Solve each inequality. Graph the solution set on a number line.

1. $-10 < 3x + 2 \le 14$

−8 −6 −4 −2 0 2 4 6 8

2. $3a + 8 < 23$ or $\frac{1}{4}a - 6 > 7$

−10 0 10 20 30 40 50 60 70

3. $18 < 4x - 10 < 50$

3 5 7 9 11 13 15 17 19

4. $5k + 2 < -13$ or $8k - 1 > 19$

−4 −3 −2 −1 0 1 2 3 4

5. $100 \le 5y - 45 \le 225$

0 10 20 30 40 50 60 70 80

6. $\frac{2}{3}b - 2 > 10$ or $\frac{3}{4}b + 5 < -4$

−24 −12 0 12 24

7. $22 < 6w - 2 < 82$

0 2 4 6 8 10 12 14 16

8. $4d - 1 > -9$ or $2d + 5 < 11$

−4 −3 −2 −1 0 1 2 3 4

1-6 Study Guide and Intervention (continued)

Solving Compound and Absolute Value Inequalities

Absolute Value Inequalities Use the definition of absolute value to rewrite an absolute value inequality as a compound inequality.

For all real numbers a and b, $b > 0$, the following statements are true.

1. If $|a| < b$, then $-b < a < b$.
2. If $|a| > b$, then $a > b$ or $a < -b$.

These statements are also true for $\leq$ and $\geq$, respectively.

Example 1 Solve $|x + 2| > 4$. Graph the solution set on a number line.

By statement 2 above, if $|x + 2| > 4$, then $x + 2 > 4$ or $x + 2 < -4$. Subtracting 2 from both sides of each inequality gives $x > 2$ or $x < -6$.

Example 2 Solve $|2x - 1| < 5$. Graph the solution set on a number line.

By statement 1 above, if $|2x - 1| < 5$, then $-5 < 2x - 1 < 5$. Adding 1 to all three parts of the inequality gives $-4 < 2x < 6$. Dividing by 2 gives $-2 < x < 3$.

Exercises

Solve each inequality. Graph the solution set on a number line.

1. $|3x + 4| < 8$

2. $|4k| + 1 > 27$

3. $\left|\dfrac{c}{2} - 3\right| \leq 5$

4. $|a + 9| \geq 30$

5. $|2f - 11| > 9$

6. $|5w + 2| < 28$

7. $|10 - 2k| < 2$

8. $\left|\dfrac{x}{2} - 5\right| + 2 > 10$

9. $|4b - 11| < 17$

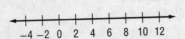

10. $|100 - 3m| > 20$

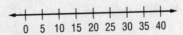

2-1 Study Guide and Intervention

Relations and Functions

Relations and Functions A **relation** can be represented as a set of ordered pairs or as an equation; the relation is then the set of all ordered pairs (x, y) that make the equation true. A **function** is a relation in which each element of the domain is paired with exactly one element of the range.

One-to-One Function	Each element of the domain pairs to exactly one **unique** element of the range.	
Onto Function	Each element of the range also corresponds to an element of the domain.	
Both One-to-One and Onto	Each element of the domain is paired to exactly one element of the range and each element of the range.	

Example State the domain and range of the relation. Does the relation represent a function?

The domain and range are both all real numbers. Each element of the domain corresponds with exactly one element of the range, so it is a function.

x	y
−1	−5
0	−3
1	−1
2	1
3	3

Exercises

State the domain and range of each relation. Then determine whether each relation is a *function*. If it is a function, determine if it is *one-to-one, onto, both,* or *neither*.

1. {(0.5, 3), (0.4, 2), (3.1, 1), (0.4, 0)}

2. {(−5, 2), (4, −2), (3, −11), (−7, 2)}

3. {(0.5, −3), (0.1, 12), (6, 8)}

4. {(−15, 12), (−14, 11), (−13, 10), (−12, 12)}

2-1 Study Guide and Intervention *(continued)*

Relations and Functions

Equations of Relations and Functions Equations that represent functions are often written in **functional notation**. For example, $y = 10 - 8x$ can be written as $f(x) = 10 - 8x$. This notation emphasizes the fact that the values of y, the **dependent variable**, depend on the values of x, the **independent variable**.

To evaluate a function, or find a functional value, means to substitute a given value in the domain into the equation to find the corresponding element in the range.

Example Given $f(x) = x^2 + 2x$, find each value.

a. $f(3)$

$f(x) = x^2 + 2x$ Original function

$f(3) = 3^2 + 2(3)$ Substitute.

$\quad\quad = 15$ Simplify.

b. $f(5a)$

$f(x) = x^2 + 2x$ Original function

$f(5a) = (5a)^2 + 2(5a)$ Substitute.

$\quad\quad = 25a^2 + 10a$ Simplify.

Exercises

Graph each relation or equation and determine the domain and range. Determine whether the relation is a *function*, is *one-to-one*, *onto*, *both*, or *neither*. Then state whether it is *discrete* or *continuous*.

1. $y = 3$

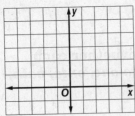

2. $y = x^2 - 1$

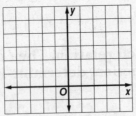

3. $y = 3x + 2$

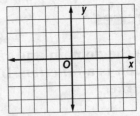

Find each value if $f(x) = -2x + 4$.

4. $f(12)$

5. $f(6)$

6. $f(2b)$

Find each value if $g(x) = x^3 - x$.

7. $g(5)$

8. $g(-2)$

9. $g(7c)$

2-2 Study Guide and Intervention
Linear Relations and Functions

Linear Relations and Functions A **linear equation** has no operations other than addition, subtraction, and multiplication of a variable by a constant. The variables may not be multiplied together or appear in a denominator. A linear equation does not contain variables with exponents other than 1. The graph of a linear equation is always a line.

A **linear function** is a function with ordered pairs that satisfy a linear equation. Any linear function can be written in the form $f(x) = mx + b$, where m and b are real numbers.

If an equation is linear, you need only two points that satisfy the equation in order to graph the equation. One way is to find the x-intercept and the y-intercept and connect these two points with a line.

Example 1 Is $f(x) = 0.2 - \frac{x}{5}$ a linear function? Explain.

Yes; it is a linear function because it can be written in the form
$f(x) = -\frac{1}{5}x + 0.2$.

Example 2 Is $2x + xy - 3y = 0$ a linear function? Explain.

No; it is not a linear function because the variables x and y are multiplied together in the middle term.

Exercises

State whether each function is a linear function. Write *yes* or *no*. Explain.

1. $6y - x = 7$

2. $9x = \frac{18}{y}$

3. $f(x) = 2 - \frac{x}{11}$

4. $2y - \frac{x}{6} - 4 = 0$

5. $1.6x - 2.4y = 4$

6. $0.2x = 100 - \frac{0.4}{y}$

7. $f(x) = 4 - x^3$

8. $f(x) = \frac{4}{x}$

9. $2yx - 3y + 2x = 0$

2-2 Study Guide and Intervention (continued)

Linear Relations and Functions

Standard Form The **standard form** of a linear equation is $Ax + By = C$, where A, B, and C are integers whose greatest common factor is 1.

Example 1 Write each equation in standard form. Identify A, B, and C.

a. $y = 8x - 5$

$y = 8x - 5$	Original equation
$-8x + y = -5$	Subtract 8x from each side.
$8x - y = 5$	Multiply each side by −1.

So $A = 8$, $B = -1$, and $C = 5$.

b. $14x = -7y + 21$

$14x = -7y + 21$	Original equation
$14x + 7y = 21$	Add 7y to each side.
$2x + y = 3$	Divide each side by 7.

So $A = 2$, $B = 1$, and $C = 3$.

Example 2 Find the x-intercept and the y-intercept of the graph of $4x - 5y = 20$. Then graph the equation.

The x-intercept is the value of x when $y = 0$.

$4x - 5y = 20$	Original equation
$4x - 5(0) = 20$	Substitute 0 for y.
$x = 5$	Simplify.

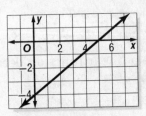

So the x-intercept is 5. Similarly, the y-intercept is −4.

Exercises

Write each equation in standard form. Identify A, B, and C.

1. $2x = 4y - 1$

2. $5y = 2x + 3$

3. $3x = -5y + 2$

4. $18y = 24x - 9$

5. $\frac{3}{4}y = \frac{2}{3}x + 5$

6. $6y - 8x + 10 = 0$

7. $0.4x + 3y = 10$

8. $x = 4y - 7$

9. $2y = 3x + 6$

Find the x-intercept and the y-intercept of the graph of each equation. Then graph the equation using the intercepts.

10. $2x + 7y = 14$

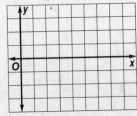

11. $5y - x = 10$

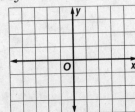

12. $2.5x - 5y + 7.5 = 0$

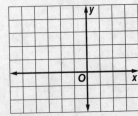

2-3 Study Guide and Intervention

Rate of Change and Slope

Rate of Change Rate of change is a ratio that compares how much one quantity changes, on average, relative to the change in another quantity.

Example Find the average rate of change for the data in the table.

$$\text{Average Rate of Change} = \frac{\text{change in } y}{\text{change in } x}$$

$$= \frac{\text{change in Elevation of the Sun}}{\text{change in Time}}$$

$$= \frac{84° - 6°}{11:00 \text{ A.M.} - 7:00 \text{ A.M.}}$$

$$= \frac{78°}{4 \text{ hours}}$$

$$= 19.5 \text{ degrees per hour}$$

Elevation of the Sun (in degrees)	Time
6°	7:00 A.M.
26°	8:00 A.M.
45°	9:00 A.M.
64°	10:00 A.M.
84°	11:00 A.M.

Exercises

Find the rate of change for each set of data.

1.

Time P.M.	People in auditorium
7:15	26
7:22	61
7:24	71
7:30	101
7:40	151

2.

Time (minutes)	Altitude of balloon (meters)
3	520
8	1,220
11	1,640
15	2,200
23	3,320

3.

Time (minutes)	Vehicles through tunnel
4	1,610
11	2,131
19	2,746
22	2,970
28	3,432

4.

Time (seconds)	Depth of sinking stone (meters)
0	3.51
7	4.77
11	5.49
21	7.29
29	8.73

5.

Time (seconds)	Water through Channel (liters)
6	22,172
13	24,706
15	25,430
23	28,326
47	37,014

6.

Time (seconds)	Distance between Two Sleds (meters)
0	37.3
3	30.2
4	27.7
7	20.8
13	7.2

2-3 Study Guide and Intervention *(continued)*

Rate of Change and Slope

Slope

Slope *m* of a Line	For points (x_1, y_1) and (x_2, y_2), where $x_1 \neq x_2$, $m = \dfrac{\text{change in } y}{\text{change in } x} = \dfrac{y_2 - y_1}{x_2 - x_1}$

Example 1 Find the slope of the line that passes through $(2, -1)$ and $(-4, 5)$.

$m = \dfrac{y_2 - y_1}{x_2 - x_1}$ Slope formula

$= \dfrac{5 - (-1)}{-4 - 2}$ $(x_1, y_1) = (2, -1), (x_2, y_2) = (-4, 5)$

$= \dfrac{6}{-6} = -1$ Simplify.

The slope of the line is -1.

Example 2 Find the slope of the line.

Find two points on the line with integer coordinates, such as $(1, -2)$ and $(3, -3)$. Divide the difference in the y–coordinates by the difference in the x–coordinates:

$\dfrac{-3 - (-2)}{3 - 1} = -\dfrac{1}{2}$

The slope of the line is $-\dfrac{1}{2}$.

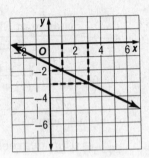

Exercises

Find the slope of the line that passes through each pair of points. Express as a fraction in simplest form.

1. $(4, 7)$ and $(6, 13)$ **2.** $(6, 4)$ and $(3, 4)$ **3.** $(5, 1)$ and $(7, -3)$

4. $(5, -3)$ and $(-4, 3)$ **5.** $(5, 10)$ and $(-1, -2)$ **6.** $(-1, -4)$ and $(-13, 2)$

7. $(7, -2)$ and $(3, 3)$ **8.** $(-5, 9)$ and $(5, 5)$ **9.** $(4, -2)$ and $(-4, -8)$

Determine the rate of change of each graph.

10.
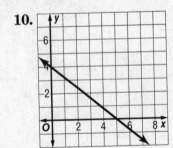

11.
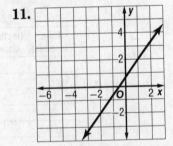

12.
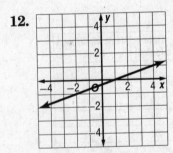

13.
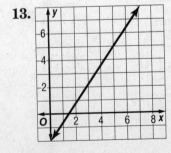

14.
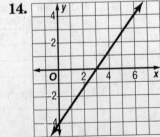

15.
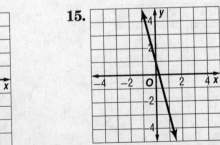

2-4 Study Guide and Intervention
Writing Linear Equations

Forms of Equations

Slope-Intercept Form of a Linear Equation	$y = mx + b$, where m is the slope and b is the y-intercept
Point-Slope Form of a Linear Equation	$y - y_1 = m(x - x_1)$, where (x_1, y_1) are the coordinates of a point on the line and m is the slope of the line

Example 1 Write an equation in slope-intercept form for the line that has slope −2 and passes through the point (3, 7).

Substitute for m, x, and y in the slope-intercept form.

$y = mx + b$	Slope-intercept form
$7 = (-2)(3) + b$	$(x, y) = (3, 7)$, $m = -2$
$7 = -6 + b$	Simplify.
$13 = b$	Add 6 to both sides.

The y-intercept is 13. The equation in slope-intercept form is $y = -2x + 13$.

Example 2 Write an equation in slope-intercept form for the line that has slope $\frac{1}{3}$ and x-intercept 5.

$y = mx + b$	Slope-intercept form
$0 = \left(\frac{1}{3}\right)(5) + b$	$(x, y) = (5, 0)$, $m = \frac{1}{3}$
$0 = \frac{5}{3} + b$	Simplify.
$-\frac{5}{3} = b$	Subtract $\frac{5}{3}$ from both sides.

The y-intercept is $-\frac{5}{3}$. The slope-intercept form is $y = \frac{1}{3}x - \frac{5}{3}$.

Exercises

Write an equation in slope-intercept form for the line described.

1. slope −2, passes through (−4, 6)

2. slope $\frac{3}{2}$, y-intercept 4

3. slope 1, passes through (2, 5)

4. slope $-\frac{13}{5}$, passes through (5, −7)

Write an equation in slope-intercept form for each graph.

5.
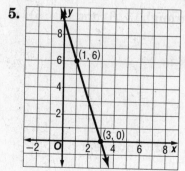

6.
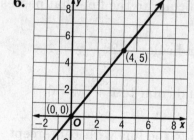

7.
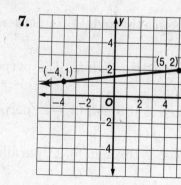

2-4 Study Guide and Intervention (continued)

Writing Linear Equations

Parallel and Perpendicular Lines Use the slope-intercept or point-slope form to find equations of lines that are parallel or perpendicular to a given line. Remember that parallel lines have equal slope. The slopes of two perpendicular lines are negative reciprocals, that is, their product is −1.

Example 1 Write an equation of the line that passes through (8, 2) and is perpendicular to the line whose equation is $y = -\frac{1}{2}x + 3$.

The slope of the given line is $-\frac{1}{2}$. Since the slopes of perpendicular lines are negative reciprocals, the slope of the perpendicular line is 2.

Use the slope and the given point to write the equation.

$y - y_1 = m(x - x_1)$ Point-slope form
$y - 2 = 2(x - 8)$ $(x_1, y_1) = (8, 2), m = 2$
$y - 2 = 2x - 16$ Distributive Prop.
$y = 2x - 14$ Add 2 to each side.

An equation of the line is $y = 2x - 14$.

Example 2 Write an equation of the line that passes through (−1, 5) and is parallel to the graph of $y = 3x + 1$.

The slope of the given line is 3. Since the slopes of parallel lines are equal, the slope of the parallel line is also 3.

Use the slope and the given point to write the equation.

$y - y_1 = m(x - x_1)$ Point-slope form
$y - 5 = 3(x - (-1))$ $(x_1, y_1) = (-1, 5), m = 3$
$y - 5 = 3x + 3$ Distributive Prop.
$y = 3x + 8$ Add 5 to each side.

An equation of the line is $y = 3x + 8$.

Exercises

Write an equation in slope-intercept form for the line that satisfies each set of conditions.

1. passes through (−4, 2), parallel to $y = \frac{1}{2}x + 5$

2. passes through (3, 1), perpendicular to $y = -3x + 2$

3. passes through (1, −1), parallel to the line that passes through (4, 1) and (2, −3)

4. passes through (4, 7), perpendicular to the line that passes through (3, 6) and (3, 15)

5. passes through (8, −6), perpendicular to $2x - y = 4$

6. passes through (2, −2), perpendicular to $x + 5y = 6$

7. passes through (6, 1), parallel to the line with x-intercept −3 and y-intercept 5

8. passes through (−2, 1), perpendicular to $y = 4x - 11$

2-5 Study Guide and Intervention

Scatter Plots and Lines of Regression

Scatter Plots and Prediction Equations A set of data points graphed as ordered pairs in a coordinate plane is called a **scatter plot**. A scatter plot can be used to determine if there is a relationship among the data. A **line of fit** is a line that closely approximates a set of data graphed in a scatter plot. The equation of a line of fit is called a **prediction equation** because it can be used to predict values not given in the data set.

Example **STORAGE COSTS** According to a certain prediction equation, the cost of 200 square feet of storage space is $60. The cost of 325 square feet of storage space is $160.

a. Find the slope of the prediction equation. What does it represent?

Since the cost depends upon the square footage, let x represent the amount of storage space in square feet and y represent the cost in dollars. The slope can be found using the formula $m = \dfrac{y_2 - y_1}{x_2 - x_1}$. So, $m = \dfrac{160 - 60}{325 - 200} = \dfrac{100}{125} = 0.8$

The slope of the prediction equation is 0.8. This means that the price of storage increases 80¢ for each one-square-foot increase in storage space.

b. Find a prediction equation.

Using the slope and one of the points on the line, you can use the point-slope form to find a prediction equation.

$y - y_1 = m(x - x_1)$ Point-slope form
$y - 60 = 0.8(x - 200)$ $(x_1, y_1) = (200, 60)$, $m = 0.8$
$y - 60 = 0.8x - 160$ Distributive Property
$y = 0.8x - 100$ Add 60 to both sides.

A prediction equation is $y = 0.8x - 100$.

Exercises

1. **SALARIES** The table below shows the years of experience for eight technicians at Lewis Techomatic and the hourly rate of pay each technician earns.

Experience (years)	9	4	3	1	10	6	12	8
Hourly Rate of Pay	$17	$10	$10	$7	$19	$12	$20	$15

a. Draw a scatter plot to show how years of experience are related to hourly rate of pay. Draw a line of fit and describe the correlation.

b. Write a prediction equation to show how years of experience (x) are related to hourly rate of pay (y).

c. Use the function to predict the hourly rate of pay for 15 years of experience.

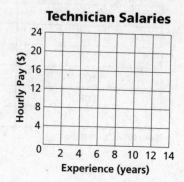

Technician Salaries

2-5 Study Guide and Intervention *(continued)*

Scatter Plots and Lines of Regression

Lines of Regression Another method for writing a line of fit is to use a line of regression. A **regression line** is determined through complex calculations to ensure that the distance of all the data points to the line of fit are at the minimum.

Example **WORLD POPULATION** The following table gives the United Nations estimates of the world population (in billions) every five years from 1980-2005. Find the equation and graph the line of regression. Then predict the population in 2010.

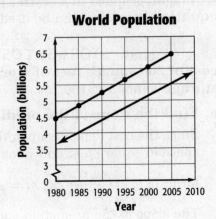

World Population

Year	Population (billions)
1980	4.451
1985	4.855
1990	5.295
1995	5.719
2000	6.124
2005	6.515
2010	?

Source: UN 2006 Revisions Population database

Step 1 Use your calculator to make a scatter plot.

Step 2 Find the equation of the line of regression.

Step 3 Graph the regression equation.

Step 4 Predict using the function.

In 2010 the population will be approximately 6.851 billion

Exercises

1. The table below shows the number of women who served in the United States Congress during the years 1995–2006. Find an equation for and graph a line of regression Then use the function to predict the number of women in Congress in the 112th Congressional Session.

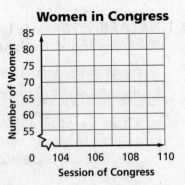

Women in Congress

Congressional Session	Number of Women
104	59
105	65
106	67
107	75
108	77
109	83

Source: U. S. Senate

2-6 Study Guide and Intervention

Special Functions

Piecewise-Defined Functions A **piecewise-defined function** is written using two or more expressions. Its graph is often disjointed.

> **Example**
>
> Graph $f(x) = \begin{cases} 2x & \text{if } x < 2 \\ x - 1 & \text{if } x \geq 2. \end{cases}$

First, graph the linear function $f(x) = 2x$ for $x < 2$. Since 2 does not satisfy this inequality, stop with a circle at (2, 4). Next, graph the linear function $f(x) = x - 1$ for $x \geq 2$. Since 2 does satisfy this inequality, begin with a dot at (2, 1).

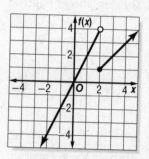

Exercises

Graph each function. Identify the domain and range.

1. $f(x) = \begin{cases} x + 2 & \text{if } x < 0 \\ 2x + 5 & \text{if } 0 \leq x \leq 2 \\ -x + 1 & \text{if } x > 2 \end{cases}$

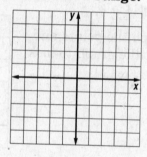

2. $f(x) = \begin{cases} -x - 4 & \text{if } x < -7 \\ 5x - 1 & \text{if } -7 \leq x \leq 0 \\ 2x + 1 & \text{if } x > 0 \end{cases}$

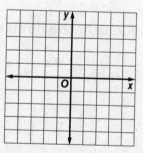

3. $h(x) = \begin{cases} \dfrac{x}{3} & \text{if } x \leq 0 \\ 2x - 6 & \text{if } 0 < x < 2 \\ 1 & \text{if } x \geq 2 \end{cases}$

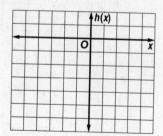

2-6 Study Guide and Intervention *(continued)*

Special Functions

Step Functions and Absolute Value Functions

Name	Written as	Graphed as
Greatest Integer Function	$f(x) = [\![x]\!]$	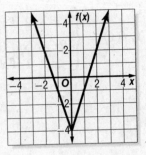
Absolute Value Function	$f(x) = \lvert x \rvert$	two rays that are mirror images of each other and meet at a point, the vertex

Example **Graph $f(x) = 3\lvert x \rvert - 4$.**

Find several ordered pairs. Graph the points and connect them. You would expect the graph to look similar to its parent function, $f(x) = \lvert x \rvert$.

x	$3\lvert x \rvert - 4$
0	−4
1	−1
2	2
−1	−1
−2	2

Exercises

Graph each function. Identify the domain and range.

1. $f(x) = 2[\![x]\!]$

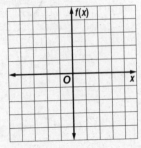

2. $h(x) = \lvert 2x + 1 \rvert$

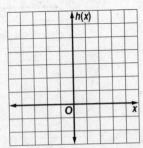

3. $f(x) = [\![x]\!] + 4$

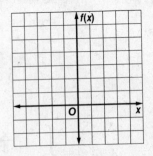

2-7 Study Guide and Intervention

Parent Functions and Transformations

Parent Graphs The **parent graph,** which is the graph of the **parent function,** is the simplest of the graphs in a family. Each graph in a **family of graphs** has similar characteristics.

Name	Characteristics	Parent Function
Constant Function	Straight horizontal line	$y = a$, where a is a real number
Identity Function	Straight diagonal line	$y = x$
Absolute Value Function	Diagonal lines shaped like a V	$y = |x|$
Quadratic Function	Curved like a parabola	$y = x^2$

Example Identify the type of function represented by each graph.

a.

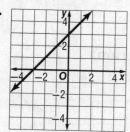

The graph is a diagonal line. The graph represents an identity function.

b.

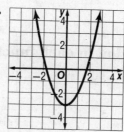

The graph is a parabolic curve. The graph represents a quadratic function.

Exercises

Identify the type of function represented by each graph.

1.

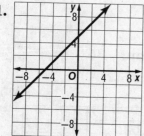

2.

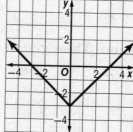

3.

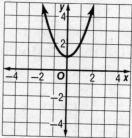

4.

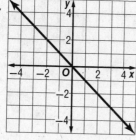

5.

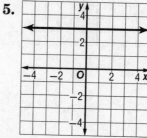

6.

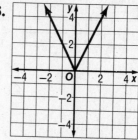

2-7 Study Guide and Intervention *(continued)*

Parent Functions and Transformations

Transformations Transformations of a parent graph may appear in a different location, may flip over an axis, or may appear to have been stretched or compressed.

Example **Describe the reflection in $y = -|x|$. Then graph the function.**

The graph of $y = -|x|$ is a reflection of the graph of $y = |x|$ in the x-axis.

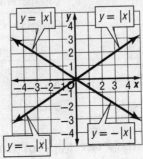

Exercises

Describe the translation in each function. Then graph the function.

1. $y = x - 4$

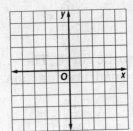

2. $y = |x + 5|$

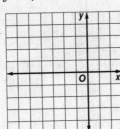

3. $y = x^2 - 3$

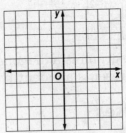

Describe the dilation in each function. Then graph the function.

4. $y = 5x$

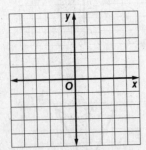

5. $y = \frac{1}{2}|x|$

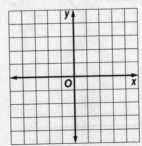

6. $y = 2x^2$

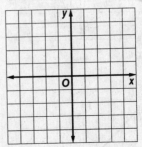

2-8 Study Guide and Intervention

Graphing Linear and Absolute Value Inequalities

Graph Linear Inequalities A **linear inequality**, like $y \geq 2x - 1$, resembles a linear equation, but with an inequality sign instead of an equals sign. The graph of the related linear equation separates the coordinate plane into two half-planes. The line is the boundary of each half-plane.

To graph a linear inequality, follow these steps.

Step 1 Graph the boundary; that is, the related linear equation. If the inequality symbol is $\leq$ or $\geq$, the boundary is solid. If the inequality symbol is $<$ or $>$, the boundary is dashed.

Step 2 Choose a point not on the boundary and test it in the inequality. $(0, 0)$ is a good point to choose if the boundary does not pass through the origin.

Step 3 If a true inequality results, shade the half-plane containing your test point. If a false inequality results, shade the other half-plane.

Example **Graph $x + 2y \geq 4$.**

The boundary is the graph of $x + 2y = 4$.

Use the slope-intercept form, $y = -\frac{1}{2}x + 2$, to graph the boundary line.

The boundary line should be solid.

Test the point $(0, 0)$.

$0 + 2(0) \overset{?}{\geq} 4$ $(x, y) = (0, 0)$

$0 \geq 4$ false

Shade the region that does *not* contain $(0, 0)$.

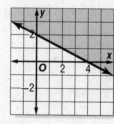

Exercises

Graph each inequality.

1. $y < 3x + 1$

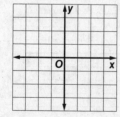

2. $y \geq x - 5$

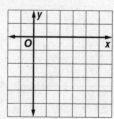

3. $4x + y \leq -1$

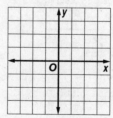

4. $y < \frac{x}{2} - 4$

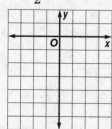

5. $x + y > 6$

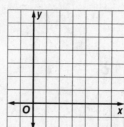

6. $0.5x - 0.25y < 1.5$

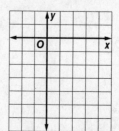

2-8 **Study Guide and Intervention** *(continued)*

Graphing Linear and Absolute Value Inequalities

Graph Absolute Value Inequalities Graphing absolute value inequalities is similar to graphing linear inequalities. The graph of the related absolute value equation is the boundary. This boundary is graphed as a solid line if the inequality is $\leq$ or $\geq$, and dashed if the inequality is $<$ or $>$. Choose a test point not on the boundary to determine which region to shade.

Example **Graph $y \leq 3|x - 1|$.**

First graph the equation $y = 3|x - 1|$.

Since the inequality is $\leq$, the graph of the boundary is solid.

Test $(0, 0)$.

$0 \overset{?}{\leq} 3|0 - 1|$ $(x, y) = (0, 0)$

$0 \overset{?}{\leq} 3|-1|$ $|-1| = 1$

$0 \leq 3$ true

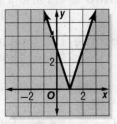

Shade the region that contains $(0, 0)$.

Exercises

Graph each inequality.

1. $y \geq |x| + 1$

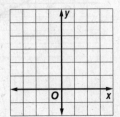

2. $y \leq |2x - 1|$

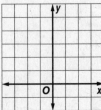

3. $y - 2|x| > 3$

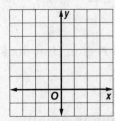

4. $y < -|x| - 3$

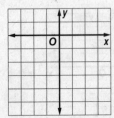

5. $|x| + y \geq 4$

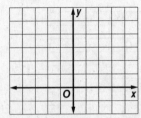

6. $|x + 1| + 2y < 0$

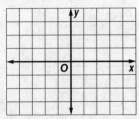

7. $|2 - x| + y > -1$

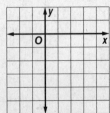

8. $y < 3|x| - 3$

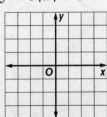

9. $y \leq |1 - x| + 4$

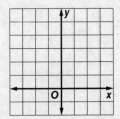

3-1 Study Guide and Intervention

Solving Systems of Equations by Graphing

Solve Systems Using Tables and Graphs A system of equations is two or more equations with the same variables. You can solve a system of linear equations by using a table or by graphing the equations on the same coordinate plane. If the lines intersect, the solution is that intersection point.

Example Solve the system of equations by graphing. $\quad x - 2y = 4$
$\quad x + y = -2$

Write each equation in slope-intercept form.

$x - 2y = 4 \quad \rightarrow \quad y = \dfrac{x}{2} - 2$

$x + y = -2 \quad \rightarrow \quad y = -x - 2$

The graphs appear to intersect at $(0, -2)$.

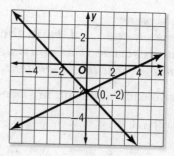

CHECK Substitute the coordinates into each equation.

$x - 2y$	$= 4$	$x + y$	$= -2$	Original equations
$0 - 2(-2)$	$\stackrel{?}{=} 4$	$0 + (-2)$	$\stackrel{?}{=} -2$	$x = 0$ and $y = -2$
	$4 = 4 \checkmark$		$-2 = -2 \checkmark$	Simplify.

The solution of the system is $(0, -2)$.

Exercises

Solve each system of equations by graphing.

1. $y = -\dfrac{x}{3} + 1$
$y = \dfrac{x}{2} - 4$

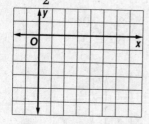

2. $y = 2x - 2$
$y = -x + 4$

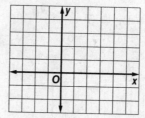

3. $y = -\dfrac{x}{2} + 3$
$y = \dfrac{x}{4}$

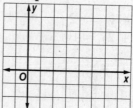

4. $3x - y = 0$
$x - y = -2$

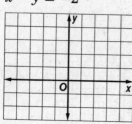

5. $2x + \dfrac{y}{3} = -7$
$\dfrac{x}{2} + y = 1$

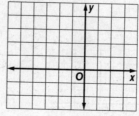

6. $\dfrac{x}{2} - y = 2$
$2x - y = -1$

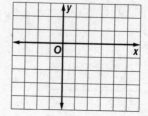

3-1 Study Guide and Intervention (continued)

Solving Systems of Equations by Graphing

Classify Systems of Equations The following chart summarizes the possibilities for graphs of two linear equations in two variables.

Graphs of Equations	Slopes of Lines	Classification of System	Number of Solutions
Lines intersect	Different slopes	Consistent and independent	One
Lines coincide (same line)	Same slope, same y-intercept	Consistent and dependent	Infinitely many
Lines are parallel	Same slope, different y-intercepts	Inconsistent	None

Example Graph the system of equations and describe it as *consistent and independent*, *consistent and dependent*, or *inconsistent*.

$$x - 3y = 6$$
$$2x - y = -3$$

Write each equation in slope-intercept form.

$x - 3y = 6 \quad \rightarrow \quad y = \frac{1}{3}x - 2$

$2x - y = -3 \quad \rightarrow \quad y = 2x + 3$

The graphs intersect at $(-3, -3)$. Since there is one solution, the system is consistent and independent.

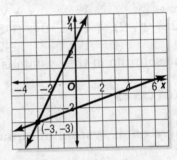

Exercises

Graph each system of equations and describe it as *consistent and independent*, *consistent and dependent*, or *inconsistent*.

1. $3x + y = -2$
  $6x + 2y = 10$

2. $x + 2y = 5$
  $3x - 15 = -6y$

3. $2x - 3y = 0$
  $4x - 6y = 3$

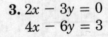

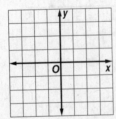

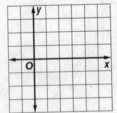

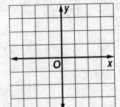

4. $2x - y = 3$
  $x + 2y = 4$

5. $4x + y = -2$
  $2x + \frac{y}{2} = -1$

6. $3x - y = 2$
  $x + y = 6$

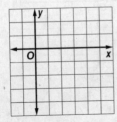

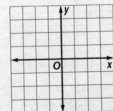

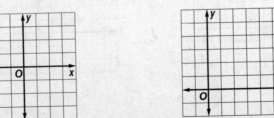

3-2 Study Guide and Intervention

Solving Systems of Equations Algebraically

Substitution To solve a system of linear equations by **substitution**, first solve for one variable in terms of the other in one of the equations. Then substitute this expression into the other equation and simplify.

Example   Use substitution to solve the system of equations. $2x - y = 9$
$x + 3y = -6$

Solve the first equation for y in terms of x.

$2x - y = 9$	First equation
$-y = -2x + 9$	Subtract 2x from both sides.
$y = 2x - 9$	Multiply both sides by -1.

Substitute the expression $2x - 9$ for y into the second equation and solve for x.

$x + 3y = -6$	Second equation
$x + 3(2x - 9) = -6$	Substitute $2x - 9$ for y.
$x + 6x - 27 = -6$	Distributive Property
$7x - 27 = -6$	Simplify.
$7x = 21$	Add 27 to each side.
$x = 3$	Divide each side by 7.

Now, substitute the value 3 for x in either original equation and solve for y.

$2x - y = 9$	First equation
$2(3) - y = 9$	Replace x with 3.
$6 - y = 9$	Simplify.
$-y = 3$	Subtract 6 from each side.
$y = -3$	Multiply each side by -1.

The solution of the system is $(3, -3)$.

Exercises

Solve each system of equations by using substitution.

1. $3x + y = 7$
$4x + 2y = 16$

2. $2x + y = 5$
$3x - 3y = 3$

3. $2x + 3y = -3$
$x + 2y = 2$

4. $2x - y = 7$
$6x - 3y = 14$

5. $4x - 3y = 4$
$2x + y = -8$

6. $5x + y = 6$
$3 - x = 0$

7. $x + 8y = -2$
$x - 3y = 20$

8. $2x - y = -4$
$4x + y = 1$

9. $x - y = -2$
$2x - 3y = 2$

10. $x - 4y = 4$
$2x + 12y = 13$

11. $x + 3y = 2$
$4x + 12y = 8$

12. $2x + 2y = 4$
$x - 2y = 0$

3-2 Study Guide and Intervention (continued)

Solving Systems of Equations Algebraically

Elimination To solve a system of linear equations by **elimination**, add or subtract the equations to eliminate one of the variables. You may first need to multiply one or both of the equations by a constant so that one of the variables has the opposite coefficient in one equation as it has in the other.

Example 1 Use the elimination method to solve the system of equations.

$2x - 4y = -26$
$3x - y = -24$

Multiply the second equation by -4. Then add the equations to eliminate the y variable.

$2x - 4y = -26$
$3x - y = -24$ ⟶ Multiply by -4.

$$\begin{array}{rcr} 2x - 4y &=& -26 \\ -12x + 4y &=& 96 \\ \hline -10x &=& 70 \\ x &=& -7 \end{array}$$

Replace x with -7 and solve for y.
$2x - 4y = -26$
$2(-7) - 4y = -26$
$-14 - 4y = -26$
$-4y = -12$
$y = 3$
The solution is $(-7, 3)$.

Example 2 Use the elimination method to solve the system of equations.

$3x - 2y = 4$
$5x + 3y = -25$

Multiply the first equation by 3 and the second equation by 2. Then add the equations to eliminate the y variable.

$3x - 2y = 4$ ⟶ Multiply by 3.
$5x + 3y = -25$ ⟶ Multiply by 2.

$$\begin{array}{rcr} 9x - 6y &=& 12 \\ 10x + 6y &=& -50 \\ \hline 19x &=& -38 \\ x &=& -2 \end{array}$$

Replace x with -2 and solve for y.
$3x - 2y = 4$
$3(-2) - 2y = 4$
$-6 - 2y = 4$
$-2y = 10$
$y = -5$
The solution is $(-2, -5)$.

Exercises

Solve each system of equations by using elimination.

1. $2x - y = 7$
   $3x + y = 8$

2. $x - 2y = 4$
   $-x + 6y = 12$

3. $3x + 4y = -10$
   $x - 4y = 2$

4. $3x - y = 12$
   $5x + 2y = 20$

5. $4x - y = 6$
   $2x - \dfrac{y}{2} = 4$

6. $5x + 2y = 12$
   $-6x - 2y = -14$

7. $2x + y = 8$
   $3x + \dfrac{3}{2}y = 12$

8. $7x + 2y = -1$
   $4x - 3y = -13$

9. $3x + 8y = -6$
   $x - y = 9$

10. $5x + 4y = 12$
    $7x - 6y = 40$

11. $-4x + y = -12$
    $4x + 2y = 6$

12. $5x + 2y = -8$
    $4x + 3y = 2$

3-3 Study Guide and Intervention

Solving Systems of Inequalities by Graphing

Systems of Inequalities To solve a system of inequalities, graph the inequalities in the same coordinate plane. The solution of the system is the region shaded for all of the inequalities.

Example Solve the system of inequalities.

$y \leq 2x - 1$ and $y > \dfrac{x}{3} + 2$

The solution of $y \leq 2x - 1$ is Regions 1 and 2.

The solution of $y > \dfrac{x}{3} + 2$ is Regions 1 and 3.

The intersection of these regions is Region 1, which is the solution set of the system of inequalities.

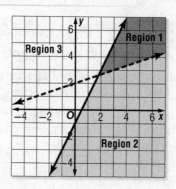

Exercises

Solve each system of inequalities by graphing.

1. $x - y \leq 2$
   $x + 2y \geq 1$

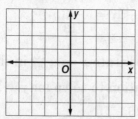

2. $3x - 2y \leq -1$
   $x + 4y \geq -12$

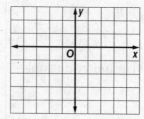

3. $|y| \leq 1$
   $x > 2$

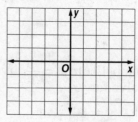

4. $y \geq \dfrac{x}{2} - 3$
   $y < 2x$

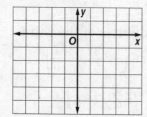

5. $y < \dfrac{x}{3} + 2$
   $y < -2x + 1$

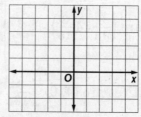

6. $y \geq -\dfrac{x}{4} + 1$
   $y < 3x - 1$

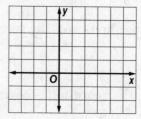

7. $x + y \geq 4$
   $2x - y > 2$

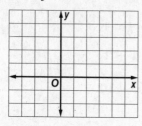

8. $x + 3y < 3$
   $x - 2y \geq 4$

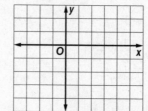

9. $x - 2y > 6$
   $x + 4y < -4$

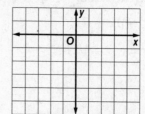

3-3 Study Guide and Intervention (continued)

Solving Systems of Inequalities by Graphing

Find Vertices of an Enclosed Region Sometimes the graph of a system of inequalities produces an enclosed region in the form of a polygon. You can find the vertices of the region by a combination of the methods used earlier in this chapter: graphing, substitution, and/or elimination.

Example **Find the coordinates of the vertices of the triangle formed by** $5x + 4y < 20$, $y < 2x + 3$, and $x - 3y < 4$.

Graph each inequality. The intersections of the boundary lines are the vertices of a triangle.

The vertex $(4, 0)$ can be determined from the graph. To find the coordinates of the second and third vertices, solve the two systems of equations

$$\begin{array}{l} y = 2x + 3 \\ 5x + 4y = 20 \end{array} \text{ and } \begin{array}{l} y = 2x + 3 \\ x - 3y = 4 \end{array}$$

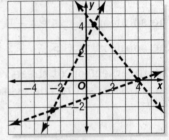

For the first system of equations, rewrite the first equation in standard form as $2x - y = -3$. Then multiply that equation by 4 and add to the second equation.

$$\begin{array}{r} 2x - y = -3 \quad \text{Multiply by 4.} \quad 8x - 4y = -12 \\ 5x + 4y = 20 \qquad\qquad (+)\ 5x + 4y = \ \ 20 \\ \hline 13x \quad = \ \ 8 \\ x \quad = \ \dfrac{8}{13} \end{array}$$

Then substitute $x = \dfrac{8}{13}$ in one of the original equations and solve for y.

$$2\left(\dfrac{8}{13}\right) - y = -3$$

$$\dfrac{16}{13} - y = -3$$

$$y = \dfrac{55}{13}$$

The coordinates of the second vertex are $\left(\dfrac{8}{13}, 4\dfrac{3}{13}\right)$.

For the second system of equations, use substitution. Substitute $2x + 3$ for y in the second equation to get

$$x - 3(2x + 3) = \quad 4$$
$$x - 6x - 9 = \quad 4$$
$$-5x = \quad 13$$
$$x = -\dfrac{13}{5}$$

Then substitute $x = -\dfrac{13}{5}$ in the first equation to solve for y.

$$y = 2\left(-\dfrac{13}{5}\right) + 3$$

$$y = -\dfrac{26}{5} + 3$$

$$y = -\dfrac{11}{5}$$

The coordinates of the third vertex are $\left(-2\dfrac{3}{5}, -2\dfrac{1}{5}\right)$.

Thus, the coordinates of the three vertices are $(4, 0)$, $\left(\dfrac{8}{13}, 4\dfrac{3}{13}\right)$ and $\left(-2\dfrac{3}{5}, -2\dfrac{1}{5}\right)$.

Exercises

Find the coordinates of the vertices of the triangle formed by each system of inequalities.

1. $y \leq -3x + 7$

$y < \dfrac{1}{2}x$

$y > -2$

2. $x > -3$

$y < -\dfrac{1}{3}x + 3$

$y > x - 1$

3. $y < -\dfrac{1}{2}x + 3$

$y > \dfrac{1}{2}x + 1$

$y < 3x + 10$

3-4 Study Guide and Intervention

Optimization with Linear Programming

Maximum and Minimum Values When a system of linear inequalities produces a bounded polygonal region, the *maximum* or *minimum* value of a related function will occur at a vertex of the region.

Example Graph the system of inequalities. Name the coordinates of the vertices of the feasible region. Find the maximum and minimum values of the function $f(x, y) = 3x + 2y$ for this polygonal region.

$$y \leq 4$$
$$y \leq -x + 6$$
$$y \geq \frac{1}{2}x - \frac{3}{2}$$
$$y \leq 6x + 4$$

First find the vertices of the bounded region. Graph the inequalities.

The polygon formed is a quadrilateral with vertices at $(0, 4)$, $(2, 4)$, $(5, 1)$, and $(-1, -2)$. Use the table to find the maximum and minimum values of $f(x, y) = 3x + 2y$.

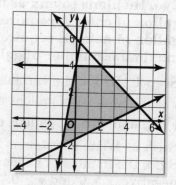

(x, y)	3x + 2y	f(x, y)
(0, 4)	3(0) + 2(4)	8
(2, 4)	3(2) + 2(4)	14
(5, 1)	3(5) + 2(1)	17
(−1, −2)	3(−1) + 2(−2)	−7

The maximum value is 17 at $(5, 1)$. The minimum value is −7 at $(-1, -2)$.

Exercises

Graph each system of inequalities. Name the coordinates of the vertices of the feasible region. Find the maximum and minimum values of the given function for this region.

1. $y \geq 2$
$1 \leq x \leq 5$
$y \leq x + 3$
$f(x, y) = 3x - 2y$

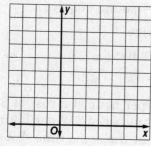

2. $y \geq -2$
$y \geq 2x - 4$
$x - 2y \geq -1$
$f(x, y) = 4x - y$

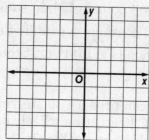

3. $x + y \geq 2$
$4y \leq x + 8$
$y \geq 2x - 5$
$f(x, y) = 4x + 3y$

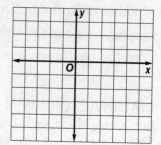

3-4 Study Guide and Intervention (continued)

Optimization with Linear Programming

Optimization When solving **linear programming** problems, use the following procedure.

1. Define variables.
2. Write a system of inequalities.
3. Graph the system of inequalities.
4. Find the coordinates of the vertices of the feasible region.
5. Write an expression to be maximized or minimized.
6. Substitute the coordinates of the vertices in the expression.
7. Select the greatest or least result to answer the problem.

Example A painter has exactly 32 units of yellow dye and 54 units of green dye. He plans to mix as many gallons as possible of color A and color B. Each gallon of color A requires 4 units of yellow dye and 1 unit of green dye. Each gallon of color B requires 1 unit of yellow dye and 6 units of green dye. Find the maximum number of gallons he can mix.

Step 1 Define the variables.

x = the number of gallons of color A made

y = the number of gallons of color B made

Step 2 Write a system of inequalities.
Since the number of gallons made cannot be negative, $x \geq 0$ and $y \geq 0$.

There are 32 units of yellow dye; each gallon of color A requires 4 units, and each gallon of color B requires 1 unit.

So $4x + y \leq 32$.

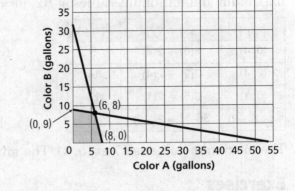

Similarly for the green dye, $x + 6y \leq 54$.

Steps 3 and 4 Graph the system of inequalities and find the coordinates of the vertices of the feasible region. The vertices of the feasible region are (0, 0), (0, 9), (6, 8), and (8, 0).

Steps 5–7 Find the maximum number of gallons, $x + y$, that he can make. The maximum number of gallons the painter can make is 14, 6 gallons of color A and 8 gallons of color B.

Exercises

1. **FOOD** A delicatessen has 12 pounds of plain sausage and 10 pounds of spicy sausage. A pound of Bratwurst A contains $\frac{3}{4}$ pound of plain sausage and $\frac{1}{4}$ pound of spicy sausage. A pound of Bratwurst B contains $\frac{1}{2}$ pound of each sausage. Find the maximum number of pounds of bratwurst that can be made.

2. **MANUFACTURING** Machine A can produce 30 steering wheels per hour at a cost of $8 per hour. Machine B can produce 40 steering wheels per hour at a cost of $12 per hour. The company can use either machine by itself or both machines at the same time. What is the minimum number of hours needed to produce 380 steering wheels if the cost must be no more than $108?

3-5 Study Guide and Intervention

Systems of Equations in Three Variables

Systems in Three Variables Use the methods used for solving systems of linear equations in two variables to solve systems of equations in three variables. A system of three equations in three variables can have a unique solution, infinitely many solutions, or no solution. A solution is an **ordered triple.**

Example Solve the system of equations.

$3x + y - z = -6$
$2x - y + 2z = 8$
$4x + y - 3z = -21$

Step 1 Use elimination to make a system of two equations in two variables.

$3x + y -$	$z = -6$	First equation		$2x - y + 2z =$	8	Second equation
$(+) 2x - y + 2z =$	8	Second equation		$(+) 4x + y - 3z =$	-21	Third equation
$5x +$	$z = 2$	Add to eliminate y.		$6x$	$- z = -13$	Add to eliminate y.

Step 2 Solve the system of two equations.

$5x + z = 2$
$(+) 6x - z = -13$
$\overline{11x = -11}$ Add to eliminate z.
$x = -1$ Divide both sides by 11.

Substitute -1 for x in one of the equations with two variables and solve for z.

$5x + z = 2$ Equation with two variables
$5(-1) + z = 2$ Replace x with −1.
$-5 + z = 2$ Multiply.
$z = 7$ Add 5 to both sides.

The result so far is $x = -1$ and $z = 7$.

Step 3 Substitute -1 for x and 7 for z in one of the original equations with three variables.

$3x + y - z = -6$ Original equation with three variables
$3(-1) + y - 7 = -6$ Replace x with −1 and z with 7.
$-3 + y - 7 = -6$ Multiply.
$y = 4$ Simplify.

The solution is $(-1, 4, 7)$.

Exercises

Solve each system of equations.

1. $2x + 3y - z = 0$
$x - 2y - 4z = 14$
$3x + y - 8z = 17$

2. $2x - y + 4z = 11$
$x + 2y - 6z = -11$
$3x - 2y - 10z = 11$

3. $x - 2y + z = 8$
$2x + y - z = 0$
$3x - 6y + 3z = 24$

4. $3x - y - z = 5$
$3x + 2y - z = 11$
$6x - 3y + 2z = -12$

5. $2x - 4y - z = 10$
$4x - 8y - 2z = 16$
$3x + y + z = 12$

6. $x - 6y + 4z = 2$
$2x + 4y - 8z = 16$
$x - 2y = 5$

3-5 Study Guide and Intervention *(continued)*

Systems of Equations in Three Variables

Real-World Problems

Example The Laredo Sports Shop sold 10 balls, 3 bats, and 2 bases for $99 on Monday. On Tuesday they sold 4 balls, 8 bats, and 2 bases for $78. On Wednesday they sold 2 balls, 3 bats, and 1 base for $33.60. What are the prices of 1 ball, 1 bat, and 1 base?

First define the variables.

x = price of 1 ball
y = price of 1 bat
z = price of 1 base

Translate the information in the problem into three equations.

$10x + 3y + 2z = 99$
$4x + 8y + 2z = 78$
$2x + 3y + z = 33.60$

Subtract the second equation from the first equation to eliminate z.

$$10x + 3y + 2z = 99$$
$$(-)\ \ 4x + 8y + 2z = 78$$
$$\overline{6x - 5y = 21}$$

Multiply the third equation by 2 and subtract from the second equation.

$$4x + 8y + 2z = 78$$
$$(-)\ 4x + 6y + 2z = 67.20$$
$$\overline{2y = 10.80}$$
$$y = 5.40$$

Substitute 5.40 for y in the equation $6x - 5y = 21$.

$$6x - 5(5.40) = 21$$
$$6x = 48$$
$$x = 8$$

Substitute 8 for x and 5.40 for y in one of the original equations to solve for z.

$$10x + 3y + 2z = 99$$
$$10(8) + 3(5.40) + 2z = 99$$
$$80 + 16.20 + 2z = 99$$
$$2z = 2.80$$
$$z = 1.40$$

So a ball costs $8, a bat $5.40, and a base $1.40.

Exercises

1. **FITNESS TRAINING** Carly is training for a triathlon. In her training routine each week, she runs 7 times as far as she swims, and she bikes 3 times as far as she runs. One week she trained a total of 232 miles. How far did she run that week?

2. **ENTERTAINMENT** At the arcade, Ryan, Sara, and Tim played video racing games, pinball, and air hockey. Ryan spent $6 for 6 racing games, 2 pinball games, and 1 game of air hockey. Sara spent $12 for 3 racing games, 4 pinball games, and 5 games of air hockey. Tim spent $12.25 for 2 racing games, 7 pinball games, and 4 games of air hockey. How much did each of the games cost?

3. **FOOD** A natural food store makes its own brand of trail mix out of dried apples, raisins, and peanuts. One pound of the mixture costs $3.18. It contains twice as much peanuts by weight as apples. One pound of dried apples costs $4.48, a pound of raisins $2.40, and a pound of peanuts $3.44. How many ounces of each ingredient are contained in 1 pound of the trail mix?

4-1 Study Guide and Intervention

Introduction to Matrices

Organize and Analyze Data

Matrix	a rectangular array of variables or constants in horizontal rows and vertical columns, usually enclosed in brackets.

A matrix can be described by its **dimensions**. A matrix with m rows and n columns is an $m \times n$ matrix.

Example 1 Owls' eggs incubate for 30 days and their fledgling period is also 30 days. Swifts' eggs incubate for 20 days and their fledgling period is 44 days. Pigeon eggs incubate for 15 days, and their fledgling period is 17 days. Eggs of the king penguin incubate for 53 days, and the fledgling time for a king penguin is 360 days. Write a 2 × 4 matrix to organize this information. **Source:** The Cambridge Factfinder

$$
\begin{array}{l}
\text{Incubation} \\
\text{Fledgling}
\end{array}
\begin{array}{cccc}
\text{Owl} & \text{Swift} & \text{Pigeon} & \text{King Penguin} \\
\end{array}
\left[
\begin{array}{cccc}
30 & 20 & 15 & 53 \\
30 & 44 & 17 & 360
\end{array}
\right]
$$

Example 2 What are the dimensions of matrix A if $A = \begin{bmatrix} 13 & 10 & -3 & 45 \\ 2 & 8 & 15 & 80 \end{bmatrix}$?

Since matrix A has 2 rows and 4 columns, the dimensions of A are 2 × 4.

Exercises

State the dimensions of each matrix.

1. $\begin{bmatrix} 15 & 5 & 27 & -4 \\ 23 & 6 & 0 & 5 \\ 14 & 70 & 24 & -3 \\ 63 & 3 & 42 & 90 \end{bmatrix}$

2. $[16 \quad 12 \quad 0]$

3. $\begin{bmatrix} 71 & 44 \\ 39 & 27 \\ 45 & 16 \\ 92 & 53 \\ 78 & 65 \end{bmatrix}$

4. A travel agent provides for potential travelers the normal high temperatures for the months of January, April, July, and October for various cities. In Boston these figures are 36°, 56°, 82°, and 63°. In Dallas they are 54°, 76°, 97°, and 79°. In Los Angeles they are 68°, 72°, 84°, and 79°. In Seattle they are 46°, 58°, 74°, and 60°. In St. Louis they are 38°, 67°, 89°, and 69°. Organize this information in a 4 × 5 matrix. **Source:** The New York Times Almanac

4-1 Study Guide and Intervention *(continued)*

Introduction to Matrices

Elements of a Matrix A matrix is a rectangular array of variables or constants in horizontal rows and vertical columns. The values are called elements and are identified by their location in the matrix. The location of an element is written as a subscript with the number of its row followed by the number of its column. For example, a_{12} is the element in the first row and second column of matrix A.

In the matrices below, 11 is the value of a_{12} in the first matrix. The value of b_{32} in the second matrix is 7.

$$A = \begin{bmatrix} 7 & 11 & 2 & 8 \\ 5 & 4 & 10 & 1 \\ 9 & 3 & 6 & 12 \end{bmatrix} \qquad B = \begin{bmatrix} 3 & 9 & 12 \\ 5 & 10 & 15 \\ 8 & 7 & 6 \\ 11 & 13 & 1 \\ 4 & 2 & 14 \end{bmatrix}$$

Example 1 Find the value of c_{23}.

$$C = \begin{bmatrix} 2 & 5 & 3 \\ 3 & 4 & 1 \end{bmatrix}$$

Since c_{23} is the element in row 2, column 3, the value of c_{23} is 1.

Example 2 Find the value of d_{54}.

$$\text{matrix } D = \begin{bmatrix} 25 & 11 & 4 & 1 & 20 \\ 7 & 8 & 9 & 12 & 13 \\ 17 & 6 & 15 & 18 & 2 \\ 22 & 16 & 21 & 24 & 19 \\ 5 & 23 & 3 & 14 & 10 \end{bmatrix}$$

Since d_{54} is the element in row 5, column 4, the value of d_{54} is 14.

Exercises

Identify each element for the following matrices.

$$F = \begin{bmatrix} 12 & 7 & 5 \\ 9 & 2 & 11 \\ 6 & 14 & 8 \\ 1 & 4 & 3 \end{bmatrix}, \qquad G = \begin{bmatrix} 1 & 14 & 13 & 12 \\ 2 & 15 & 20 & 11 \\ 3 & 16 & 19 & 10 \\ 4 & 17 & 18 & 9 \\ 5 & 6 & 7 & 8 \end{bmatrix}, \qquad H = \begin{bmatrix} 5 & 9 & 11 & 4 \\ 3 & 7 & 2 & 10 \\ 8 & 2 & 6 & 1 \end{bmatrix}.$$

1. f_{32}

2. g_{51}

3. h_{22}

4. g_{43}

5. h_{34}

6. f_{23}

7. h_{14}

8. f_{42}

9. g_{14}

4-2 Study Guide and Intervention

Operations with Matrices

Add and Subtract Matrices Matrices with the same dimensions can be added together or one can be subtracted from the other.

Addition of Matrices	$\begin{bmatrix} a & b & c \\ d & e & f \\ g & h & i \end{bmatrix} + \begin{bmatrix} j & k & l \\ m & n & o \\ p & q & r \end{bmatrix} = \begin{bmatrix} a+j & b+k & c+l \\ d+m & e+n & f+o \\ g+p & h+q & i+r \end{bmatrix}$
Subtraction of Matrices	$\begin{bmatrix} a & b & c \\ d & e & f \\ g & h & i \end{bmatrix} - \begin{bmatrix} j & k & l \\ m & n & o \\ p & q & r \end{bmatrix} = \begin{bmatrix} a-j & b-k & c-l \\ d-m & e-n & f-o \\ g-p & h-q & i-r \end{bmatrix}$

Example 1

Find $A + B$ if $A = \begin{bmatrix} 6 & -7 \\ 2 & -12 \end{bmatrix}$ and $B = \begin{bmatrix} 4 & 2 \\ -5 & -6 \end{bmatrix}$.

$A + B = \begin{bmatrix} 6 & -7 \\ 2 & -12 \end{bmatrix} + \begin{bmatrix} 4 & 2 \\ -5 & -6 \end{bmatrix}$

$= \begin{bmatrix} 6+4 & -7+2 \\ 2+(-5) & -12+(-6) \end{bmatrix}$

$= \begin{bmatrix} 10 & -5 \\ -3 & -18 \end{bmatrix}$

Example 2

Find $A - B$ if $A = \begin{bmatrix} -2 & 8 \\ 3 & -4 \\ 10 & 7 \end{bmatrix}$ and $B = \begin{bmatrix} 4 & -3 \\ -2 & 1 \\ -6 & 8 \end{bmatrix}$.

$A - B = \begin{bmatrix} -2 & 8 \\ 3 & -4 \\ 10 & 7 \end{bmatrix} - \begin{bmatrix} 4 & -3 \\ -2 & 1 \\ -6 & 8 \end{bmatrix}$

$= \begin{bmatrix} -2-4 & 8-(-3) \\ 3-(-2) & -4-1 \\ 10-(-6) & 7-8 \end{bmatrix} = \begin{bmatrix} -6 & 11 \\ 5 & -5 \\ 16 & -1 \end{bmatrix}$

Exercises

Perform the indicated operations. If the matrix does not exist, write *impossible*.

1. $\begin{bmatrix} 8 & 7 \\ -10 & -6 \end{bmatrix} - \begin{bmatrix} -4 & 3 \\ 2 & -12 \end{bmatrix}$

2. $\begin{bmatrix} 6 & -5 & 9 \\ -3 & 4 & 5 \end{bmatrix} + \begin{bmatrix} -4 & 3 & 2 \\ 6 & 9 & -4 \end{bmatrix}$

3. $\begin{bmatrix} 6 \\ -3 \\ 2 \end{bmatrix} + \begin{bmatrix} -6 & 3 & -2 \end{bmatrix}$

4. $\begin{bmatrix} 5 & -2 \\ -4 & 6 \\ 7 & 9 \end{bmatrix} + \begin{bmatrix} -11 & 6 \\ 2 & -5 \\ -4 & -7 \end{bmatrix}$

5. $\begin{bmatrix} 8 & 0 & -6 \\ 4 & 5 & -11 \\ -7 & 3 & 4 \end{bmatrix} - \begin{bmatrix} -2 & 1 & 7 \\ 3 & -4 & 3 \\ -8 & 5 & 6 \end{bmatrix}$

6. $\begin{bmatrix} \frac{3}{4} & \frac{2}{5} \\ -\frac{1}{2} & \frac{4}{3} \end{bmatrix} - \begin{bmatrix} \frac{1}{2} & \frac{2}{3} \\ \frac{2}{3} & -\frac{1}{2} \end{bmatrix}$

4-2 Study Guide and Intervention (continued)

Operations with Matrices

Scalar Multiplication You can multiply an $m \times n$ matrix by a scalar k.

Scalar Multiplication	$k \begin{bmatrix} a & b & c \\ d & e & f \end{bmatrix} = \begin{bmatrix} ka & kb & kc \\ kd & ke & kf \end{bmatrix}$

Example If $A = \begin{bmatrix} 4 & 0 \\ -6 & 3 \end{bmatrix}$ and $B = \begin{bmatrix} -1 & 5 \\ 7 & 8 \end{bmatrix}$, find $3B - 2A$.

$3B - 2A = 3 \begin{bmatrix} -1 & 5 \\ 7 & 8 \end{bmatrix} - 2 \begin{bmatrix} 4 & 0 \\ -6 & 3 \end{bmatrix}$ Substitution

$= \begin{bmatrix} 3(-1) & 3(5) \\ 3(7) & 3(8) \end{bmatrix} - \begin{bmatrix} 2(4) & 2(0) \\ 2(-6) & 2(3) \end{bmatrix}$ Multiply.

$= \begin{bmatrix} -3 & 15 \\ 21 & 24 \end{bmatrix} - \begin{bmatrix} 8 & 0 \\ -12 & 6 \end{bmatrix}$ Simplify.

$= \begin{bmatrix} -3-8 & 15-0 \\ 21-(-12) & 24-6 \end{bmatrix}$ Subtract.

$= \begin{bmatrix} -11 & 15 \\ 33 & 18 \end{bmatrix}$ Simplify.

Exercises

Perform the indicated operations. If the matrix does not exist, write *impossible*.

1. $6 \begin{bmatrix} 2 & -5 & 3 \\ 0 & 7 & -1 \\ -4 & 6 & 9 \end{bmatrix}$

2. $-\dfrac{1}{3} \begin{bmatrix} 6 & 15 & 9 \\ 51 & -33 & 24 \\ -18 & 3 & 45 \end{bmatrix}$

3. $0.2 \begin{bmatrix} 25 & -10 & -45 \\ 5 & 55 & -30 \\ 60 & 35 & -95 \end{bmatrix}$

4. $3 \begin{bmatrix} -4 & 5 \\ 2 & 3 \end{bmatrix} - 2 \begin{bmatrix} -1 & 2 \\ -3 & 5 \end{bmatrix}$

5. $-2 \begin{bmatrix} 3 & -1 \\ 0 & 7 \end{bmatrix} + 4 \begin{bmatrix} -2 & 0 \\ 2 & 5 \end{bmatrix}$

6. $2 \begin{bmatrix} 6 & -10 \\ -5 & 8 \end{bmatrix} + 5 \begin{bmatrix} 2 & 1 \\ 4 & 3 \end{bmatrix}$

7. $4 \begin{bmatrix} 1 & -2 & 5 \\ -3 & 4 & 1 \end{bmatrix} - 2 \begin{bmatrix} 4 & 3 & -4 \\ 2 & -5 & -1 \end{bmatrix}$

8. $8 \begin{bmatrix} 2 & 1 \\ 3 & -1 \\ -2 & 4 \end{bmatrix} + 3 \begin{bmatrix} 4 & 0 \\ -2 & 3 \\ 3 & -4 \end{bmatrix}$

9. $\dfrac{1}{4} \left(\begin{bmatrix} 9 & 1 \\ -7 & 0 \end{bmatrix} + \begin{bmatrix} 3 & -5 \\ 1 & 7 \end{bmatrix} \right)$

4-3 Study Guide and Intervention

Multiplying Matrices

Multiply Matrices You can multiply two matrices if and only if the number of columns in the first matrix is equal to the number of rows in the second matrix.

Multiplication of Matrices	$A \cdot B = AB$ $\begin{bmatrix} a & b \\ c & d \end{bmatrix} \cdot \begin{bmatrix} e & f \\ g & h \end{bmatrix} = \begin{bmatrix} ae + bg & af + bh \\ ce + dg & cf + dh \end{bmatrix}$

Example

Find AB if $A = \begin{bmatrix} -4 & 3 \\ 2 & -2 \\ 1 & 7 \end{bmatrix}$ and $B = \begin{bmatrix} 5 & -2 \\ -1 & 3 \end{bmatrix}$.

$AB = \begin{bmatrix} -4 & 3 \\ 2 & -2 \\ 1 & 7 \end{bmatrix} \cdot \begin{bmatrix} 5 & -2 \\ -1 & 3 \end{bmatrix}$ Substitution

$= \begin{bmatrix} -4(5) + 3(-1) & -4(-2) + 3(3) \\ 2(5) + (-2)(-1) & 2(-2) + (-2)(3) \\ 1(5) + 7(-1) & 1(-2) + 7(3) \end{bmatrix}$ Multiply columns by rows.

$= \begin{bmatrix} -23 & 17 \\ 12 & -10 \\ -2 & 19 \end{bmatrix}$ Simplify.

Exercises

Find each product, if possible.

1. $\begin{bmatrix} 4 & 1 \\ -2 & 3 \end{bmatrix} \cdot \begin{bmatrix} 3 & 0 \\ 0 & 3 \end{bmatrix}$

2. $\begin{bmatrix} -1 & 0 \\ 3 & 7 \end{bmatrix} \cdot \begin{bmatrix} 3 & 2 \\ -1 & 4 \end{bmatrix}$

3. $\begin{bmatrix} 3 & -1 \\ 2 & 4 \end{bmatrix} \cdot \begin{bmatrix} 3 & -1 \\ 2 & 4 \end{bmatrix}$

4. $\begin{bmatrix} -3 & 1 \\ 5 & -2 \end{bmatrix} \cdot \begin{bmatrix} 4 & 0 & -2 \\ -3 & 1 & 1 \end{bmatrix}$

5. $\begin{bmatrix} 3 & -2 \\ 0 & 4 \\ -5 & 1 \end{bmatrix} \cdot \begin{bmatrix} 1 & 2 \\ 2 & 1 \end{bmatrix}$

6. $\begin{bmatrix} 5 & -2 \\ 2 & -3 \end{bmatrix} \cdot \begin{bmatrix} 4 & -1 \\ -2 & 5 \end{bmatrix}$

7. $\begin{bmatrix} 6 & 10 \\ -4 & 3 \\ -2 & 7 \end{bmatrix} \cdot [0 \quad 4 \quad -3]$

8. $\begin{bmatrix} 7 & -2 \\ 5 & -4 \end{bmatrix} \cdot \begin{bmatrix} 1 & -3 \\ -2 & 0 \end{bmatrix}$

9. $\begin{bmatrix} 2 & 0 & -3 \\ 1 & 4 & -2 \\ -1 & 3 & 1 \end{bmatrix} \cdot \begin{bmatrix} 2 & -2 \\ 3 & 1 \\ -2 & 4 \end{bmatrix}$

4-3 Study Guide and Intervention *(continued)*

Multiplying Matrices

Multiplicative Properties The Commutative Property of Multiplication does *not* hold for matrices.

Properties of Matrix Multiplication	For any matrices A, B, and C for which the matrix product is defined, and any scalar c, the following properties are true.
Associative Property of Matrix Multiplication	$(AB)C = A(BC)$
Associative Property of Scalar Multiplication	$c(AB) = (cA)B = A(cB)$
Left Distributive Property	$C(A + B) = CA + CB$
Right Distributive Property	$(A + B)C = AC + BC$

Example Use $A = \begin{bmatrix} 4 & -3 \\ 2 & 1 \end{bmatrix}$, $B = \begin{bmatrix} 2 & 0 \\ 5 & -3 \end{bmatrix}$, and $C = \begin{bmatrix} 1 & -2 \\ 6 & 3 \end{bmatrix}$ to find each product.

a. $(A + B)C$

$$(A + B)C = \left(\begin{bmatrix} 4 & -3 \\ 2 & 1 \end{bmatrix} + \begin{bmatrix} 2 & 0 \\ 5 & -3 \end{bmatrix} \right) \cdot \begin{bmatrix} 1 & -2 \\ 6 & 3 \end{bmatrix}$$

$$= \begin{bmatrix} 6 & -3 \\ 7 & -2 \end{bmatrix} \cdot \begin{bmatrix} 1 & -2 \\ 6 & 3 \end{bmatrix}$$

$$= \begin{bmatrix} 6(1) + (-3)(6) & 6(-2) + (-3)(3) \\ 7(1) + (-2)(6) & 7(-2) + (-2)(3) \end{bmatrix}$$

$$= \begin{bmatrix} -12 & -21 \\ -5 & -20 \end{bmatrix}$$

b. $AC + BC$

$$AC + BC = \begin{bmatrix} 4 & -3 \\ 2 & 1 \end{bmatrix} \cdot \begin{bmatrix} 1 & -2 \\ 6 & 3 \end{bmatrix} + \begin{bmatrix} 2 & 0 \\ 5 & -3 \end{bmatrix} \cdot \begin{bmatrix} 1 & -2 \\ 6 & 3 \end{bmatrix}$$

$$= \begin{bmatrix} 4(1) + (-3)(6) & 4(-2) + (-3)(3) \\ 2(1) + 1(6) & 2(-2) + 1(3) \end{bmatrix} + \begin{bmatrix} 2(1) + 0(6) & 2(-2) + 0(3) \\ 5(1) + (-3)(6) & 5(-2) + (-3)(3) \end{bmatrix}$$

$$= \begin{bmatrix} -14 & -17 \\ 8 & -1 \end{bmatrix} + \begin{bmatrix} 2 & -4 \\ -13 & -19 \end{bmatrix} = \begin{bmatrix} -12 & -21 \\ -5 & -20 \end{bmatrix}$$

Note that although the results in the example illustrate the Right Distributive Property, they do not prove it.

Exercises

Use $A = \begin{bmatrix} 3 & 2 \\ 5 & -2 \end{bmatrix}$, $B = \begin{bmatrix} 6 & 4 \\ 2 & 1 \end{bmatrix}$, $C = \begin{bmatrix} -\frac{1}{2} & -2 \\ 1 & -3 \end{bmatrix}$, and scalar $c = -4$ to determine whether the following equations are true for the given matrices.

1. $c(AB) = (cA)B$

2. $AB = BA$

3. $BC = CB$

4. $(AB)C = A(BC)$

5. $C(A + B) = AC + BC$

6. $c(A + B) = cA + cB$

4-4 Study Guide and Intervention

Transformations with Matrices

Translations and Dilations Matrices that represent coordinates of points on a plane are useful in describing transformations.

Translation	a transformation that moves a figure from one location to another on the coordinate plane

You can use matrix addition and a translation matrix to find the coordinates of the translated figure.

Dilation	a transformation in which a figure is enlarged or reduced

You can use scalar multiplication to perform dilations.

Example Find the coordinates of the vertices of the image of $\triangle ABC$ with vertices $A(-5, 4)$, $B(-1, 5)$, and $C(-3, -1)$ if it is moved 6 units to the right and 4 units down. Then graph $\triangle ABC$ and its image $\triangle A'B'C'$.

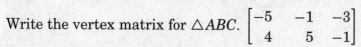

Write the vertex matrix for $\triangle ABC$. $\begin{bmatrix} -5 & -1 & -3 \\ 4 & 5 & -1 \end{bmatrix}$

Add the translation matrix $\begin{bmatrix} 6 & 6 & 6 \\ -4 & -4 & -4 \end{bmatrix}$ to the vertex matrix of $\triangle ABC$.

$$\begin{bmatrix} -5 & -1 & -3 \\ 4 & 5 & -1 \end{bmatrix} + \begin{bmatrix} 6 & 6 & 6 \\ -4 & -4 & -4 \end{bmatrix} = \begin{bmatrix} 1 & 5 & 3 \\ 0 & 1 & -5 \end{bmatrix}$$

The coordinates of the vertices of $\triangle A'B'C'$ are $A'(1, 0)$, $B'(5, 1)$, and $C'(3, -5)$.

Exercises

1. Quadrilateral $QUAD$ with vertices $Q(-1, -3)$, $U(0, 0)$, $A(5, -1)$, and $D(2, -5)$ is translated 3 units to the left and 2 units up.

   a. Write the translation matrix.

   b. Find the coordinates of the vertices of $Q'U'A'D'$.

2. The vertices of $\triangle ABC$ are $A(4, -2)$, $B(2, 8)$, and $C(8, 2)$. The triangle is dilated so that its perimeter is one-fourth the original perimeter.

   a. Write the coordinates of the vertices of $\triangle ABC$ in a vertex matrix.

   b. Find the coordinates of the vertices of image $\triangle A'B'C'$.

   c. Graph the preimage and the image.

4-4 Study Guide and Intervention *(continued)*

Transformations with Matrices

Reflections and Rotations

Reflection Matrices	For a reflection in the:	x-axis	y-axis	line y = x
	multiply the vertex matrix on the left by:	$\begin{bmatrix} 1 & 0 \\ 0 & -1 \end{bmatrix}$	$\begin{bmatrix} -1 & 0 \\ 0 & 1 \end{bmatrix}$	$\begin{bmatrix} 0 & 1 \\ 1 & 0 \end{bmatrix}$
Rotation Matrices	For a counterclockwise rotation about the origin of:	90°	180°	270°
	multiply the vertex matrix on the left by:	$\begin{bmatrix} 0 & -1 \\ 1 & 0 \end{bmatrix}$	$\begin{bmatrix} -1 & 0 \\ 0 & -1 \end{bmatrix}$	$\begin{bmatrix} 0 & 1 \\ -1 & 0 \end{bmatrix}$

Example **Find the coordinates of the vertices of the image of $\triangle ABC$ with $A(3, 5)$, $B(-2, 4)$, and $C(1, -1)$ after a reflection in the line $y = x$.**

Write the ordered pairs as a vertex matrix. Then multiply the vertex matrix by the reflection matrix for $y = x$.

$$\begin{bmatrix} 0 & 1 \\ 1 & 0 \end{bmatrix} \cdot \begin{bmatrix} 3 & -2 & 1 \\ 5 & 4 & -1 \end{bmatrix} = \begin{bmatrix} 5 & 4 & -1 \\ 3 & -2 & -1 \end{bmatrix}$$

The coordinates of the vertices of $A'B'C'$ are $A'(5, 3)$, $B'(4, -2)$, and $C'(-1, 1)$.

Exercises

1. The coordinates of the vertices of quadrilateral $ABCD$ are $A(-2, 1)$, $B(-1, 3)$, $C(2, 2)$, and $D(2, -1)$. Find the coordinates of the vertices of the image $A'B'C'D'$ after a reflection in the y-axis.

2. $\triangle DEF$ with vertices $D(-2, 5)$, $E(1, 4)$, and $F(0, -1)$ is rotated 90° counterclockwise about the origin.

   a. Write the coordinates of the triangle in a vertex matrix.

   b. Write the rotation matrix for this situation.

   c. Find the coordinates of the vertices of $\triangle D'E'F'$.

   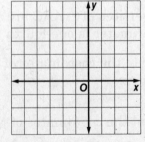

   d. Graph the preimage and the image.

Glencoe Algebra 2

4-5　Study Guide and Intervention

Determinants and Cramer's Rule

Determinants A 2×2 matrix has a second-order determinant; a 3×3 matrix has a third-order determinant.

Second-Order Determinant	For the matrix $\begin{bmatrix} a & b \\ c & d \end{bmatrix}$, the determinant is $\begin{vmatrix} a & b \\ c & d \end{vmatrix} = ad - bc$.
Third-Order Determinant	For the matrix $\begin{bmatrix} a & b & c \\ d & e & f \\ g & h & i \end{bmatrix}$, the determinant is found using the diagonal rule. 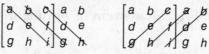
Area of a Triangle	The area of a triangle having vertices (a, b), (c, d), and (e, f) is $\lvert A \rvert$, Where $A = \dfrac{1}{2}\begin{vmatrix} a & b & 1 \\ c & d & 1 \\ e & f & 1 \end{vmatrix}$.

Example　Evaluate each determinant.

a. $\begin{vmatrix} 6 & 3 \\ -8 & 5 \end{vmatrix}$

$\begin{vmatrix} 6 & 3 \\ -8 & 5 \end{vmatrix} = 6\,(5) - 3\,(-8)$

$\qquad\qquad = 54$

b. $\begin{vmatrix} 4 & 5 & -2 \\ 1 & 3 & 0 \\ 2 & -3 & 6 \end{vmatrix}$

$\begin{vmatrix} 4 & 5 & -2 \\ 1 & 3 & 0 \\ 2 & -3 & 6 \end{vmatrix} \begin{matrix} 4 & 5 \\ 1 & 3 \\ 2 & -3 \end{matrix} \qquad \begin{vmatrix} 4 & 5 & -2 \\ 1 & 3 & 0 \\ 2 & -3 & 6 \end{vmatrix} \begin{matrix} 4 & 5 \\ 1 & 3 \\ 2 & -3 \end{matrix}$

$= [4(3)6 + 5(0)2 + (-2)1(-3)] - [(-2)3(2) + 4(0)(-3) + 5(1)6]$

$= [72 + 0 + 6] - [-12 + 0 + 30]$

$= 78 - 16 = 60$

Exercises

Evaluate each determinant.

1. $\begin{vmatrix} 6 & -2 \\ 5 & 7 \end{vmatrix}$

2. $\begin{vmatrix} 3 & 2 \\ 9 & 6 \end{vmatrix}$

3. $\begin{vmatrix} 3 & -2 & -2 \\ 0 & 4 & 1 \\ -1 & 4 & -3 \end{vmatrix}$

4. Find the area of a triangle with vertices $(2, -3)$, $(7, 4)$, and $(-5, 5)$.

4-5 Study Guide and Intervention (continued)

Determinants and Cramer's Rule

Cramer's Rule Determinants provide a way for solving systems of equations.

Cramer's Rule for Two-Variable Systems	Let C be the coefficient matrix of the system $\begin{aligned} ax + by &= m \\ fx + gy &= n \end{aligned} \rightarrow \begin{vmatrix} a & b \\ f & g \end{vmatrix}$ The solution of this system is $x = \dfrac{\begin{vmatrix} m & b \\ n & g \end{vmatrix}}{|C|}$, $y = \dfrac{\begin{vmatrix} a & m \\ f & n \end{vmatrix}}{|C|}$, if $C \neq 0$.

Example Use Cramer's Rule to solve the system of equations. $5x - 10y = 8$
$$10x + 25y = -2$$

$$x = \frac{\begin{vmatrix} m & b \\ n & g \end{vmatrix}}{|C|} \qquad \text{Cramer's Rule} \qquad y = \frac{\begin{vmatrix} a & m \\ f & n \end{vmatrix}}{|C|}$$

$$= \frac{\begin{vmatrix} 8 & -10 \\ -2 & 25 \end{vmatrix}}{\begin{vmatrix} 5 & -10 \\ 10 & 25 \end{vmatrix}} \quad a = 5, b = -10, f = 10, g = 25, m = 8, n = -2 \quad = \frac{\begin{vmatrix} 5 & 8 \\ 10 & -2 \end{vmatrix}}{\begin{vmatrix} 5 & -10 \\ 10 & 25 \end{vmatrix}}$$

$$= \frac{8(25) - (-2)(-10)}{5(25) - (-10)(10)} \qquad \text{Evaluate each determinant.} \qquad = \frac{5(-2) - 8(10)}{5(25) - (-10)(10)}$$

$$= \frac{180}{225} \text{ or } \frac{4}{5} \qquad \text{Simplify.} \qquad = -\frac{90}{225} \text{ or } -\frac{2}{5}$$

The solution is $\left(\frac{4}{5}, -\frac{2}{5}\right)$.

Exercises

Use Cramer's Rule to solve each system of equations.

1. $3x - 2y = 7$
$2x + 7y = 38$

2. $x - 4y = 17$
$3x - y = 29$

3. $2x - y = -2$
$4x - y = 4$

4. $2x - y = 1$
$5x + 2y = -29$

5. $4x + 2y = 1$
$5x - 4y = 24$

6. $6x - 3y = -3$
$2x + y = 21$

7. $2x + 7y = 16$
$x - 2y = 30$

8. $2x - 3y = -2$
$3x - 4y = 9$

9. $\dfrac{x}{3} + \dfrac{y}{5} = 2$
$\dfrac{x}{4} - \dfrac{y}{6} = -8$

10. $6x - 9y = -1$
$3x + 18y = 12$

11. $3x - 12y = -14$
$9x + 6y = -7$

12. $8x + 2y = \dfrac{3}{7}$
$5x - 4y = -\dfrac{27}{7}$

4-6 Study Guide and Intervention

Inverse Matrices and Systems of Equations

Identity and Inverse Matrices The identity matrix for matrix multiplication is a square matrix with 1s for every element of the main diagonal and zeros elsewhere.

Identity Matrix for Multiplication	If A is an $n \times n$ matrix and I is the identity matrix, then $A \cdot I = A$ and $I \cdot A = A$.

If an $n \times n$ matrix A has an inverse A^{-1}, then $A \cdot A^{-1} = A^{-1} \cdot A = I$.

Example

Determine whether $X = \begin{bmatrix} 7 & 4 \\ 10 & 6 \end{bmatrix}$ and $Y = \begin{bmatrix} 3 & -2 \\ -5 & \frac{7}{2} \end{bmatrix}$ are inverse matrices.

Find $X \cdot Y$.

$$X \cdot Y = \begin{bmatrix} 7 & 4 \\ 10 & 6 \end{bmatrix} \cdot \begin{bmatrix} 3 & -2 \\ -5 & \frac{7}{2} \end{bmatrix}$$

$$= \begin{bmatrix} 21 - 20 & -14 + 14 \\ 30 - 30 & -20 + 21 \end{bmatrix} \text{ or } \begin{bmatrix} 1 & 0 \\ 0 & 1 \end{bmatrix}$$

Find $Y \cdot X$.

$$Y \cdot X = \begin{bmatrix} 3 & -2 \\ -5 & \frac{7}{2} \end{bmatrix} \cdot \begin{bmatrix} 7 & 4 \\ 10 & 6 \end{bmatrix}$$

$$= \begin{bmatrix} 21 - 20 & 12 - 12 \\ -35 + 35 & -20 + 21 \end{bmatrix} \text{ or } \begin{bmatrix} 1 & 0 \\ 0 & 1 \end{bmatrix}$$

Since $X \cdot Y = Y \cdot X = I$, X and Y are inverse matrices.

Exercises

Determine whether the matrices in each pair are inverses of each other.

1. $\begin{bmatrix} 4 & 5 \\ 3 & 4 \end{bmatrix}$ and $\begin{bmatrix} 4 & -5 \\ -3 & 4 \end{bmatrix}$

2. $\begin{bmatrix} 3 & 2 \\ 5 & 4 \end{bmatrix}$ and $\begin{bmatrix} 2 & -1 \\ -\frac{5}{2} & \frac{3}{2} \end{bmatrix}$

3. $\begin{bmatrix} 2 & 3 \\ 5 & -1 \end{bmatrix}$ and $\begin{bmatrix} 2 & 3 \\ -1 & -2 \end{bmatrix}$

4. $\begin{bmatrix} 8 & 11 \\ 3 & 14 \end{bmatrix}$ and $\begin{bmatrix} -4 & 11 \\ 3 & -8 \end{bmatrix}$

5. $\begin{bmatrix} 4 & -1 \\ 5 & 3 \end{bmatrix}$ and $\begin{bmatrix} 1 & 2 \\ 3 & 8 \end{bmatrix}$

6. $\begin{bmatrix} 5 & 2 \\ 11 & 4 \end{bmatrix}$ and $\begin{bmatrix} -2 & 1 \\ \frac{11}{2} & -\frac{5}{2} \end{bmatrix}$

7. $\begin{bmatrix} 4 & 2 \\ 6 & -2 \end{bmatrix}$ and $\begin{bmatrix} -\frac{1}{5} & -\frac{1}{10} \\ \frac{3}{10} & \frac{1}{10} \end{bmatrix}$

8. $\begin{bmatrix} 5 & 8 \\ 4 & 6 \end{bmatrix}$ and $\begin{bmatrix} -3 & 4 \\ 2 & -\frac{5}{2} \end{bmatrix}$

9. $\begin{bmatrix} 3 & 7 \\ 2 & 4 \end{bmatrix}$ and $\begin{bmatrix} \frac{7}{2} & -\frac{3}{2} \\ 1 & -2 \end{bmatrix}$

10. $\begin{bmatrix} 3 & 2 \\ 4 & -6 \end{bmatrix}$ and $\begin{bmatrix} 3 & 2 \\ -4 & -3 \end{bmatrix}$

11. $\begin{bmatrix} 7 & 2 \\ 17 & 5 \end{bmatrix}$ and $\begin{bmatrix} 5 & -2 \\ -17 & 7 \end{bmatrix}$

12. $\begin{bmatrix} 4 & 3 \\ 7 & 5 \end{bmatrix}$ and $\begin{bmatrix} -5 & 3 \\ 7 & -4 \end{bmatrix}$

4-6 Study Guide and Intervention *(continued)*

Inverse Matrices and Systems of Equations

Matrix Equations A **matrix equation** for a system of equations consists of the product of the coefficient and variable matrices on the left and the constant matrix on the right of the equals sign.

Example Use a matrix equation to solve a system of equations.

$$3x - 7y = 12$$
$$x + 5y = -8$$

Determine the coefficient, variable, and constant matrices.

$$\begin{bmatrix} 3 & -7 \\ 1 & 5 \end{bmatrix} \cdot \begin{bmatrix} x \\ y \end{bmatrix} = \begin{bmatrix} 12 \\ -8 \end{bmatrix}$$

Find the inverse of the coefficient matrix.

$$\frac{1}{3(5) - 1(-7)} \begin{bmatrix} 5 & 7 \\ -1 & 3 \end{bmatrix} = \begin{bmatrix} \dfrac{5}{22} & \dfrac{7}{22} \\ -\dfrac{1}{22} & \dfrac{3}{22} \end{bmatrix}$$

Rewrite the equation in the form of $X = A^{-1}B$

$$\begin{bmatrix} x \\ y \end{bmatrix} = \begin{bmatrix} \dfrac{5}{22} & \dfrac{7}{22} \\ -\dfrac{1}{22} & \dfrac{3}{22} \end{bmatrix} \begin{bmatrix} 12 \\ -8 \end{bmatrix}$$

Solve.

$$\begin{bmatrix} x \\ y \end{bmatrix} = \begin{bmatrix} \dfrac{2}{11} \\ -\dfrac{18}{11} \end{bmatrix}$$

Exercises

Use a matrix equation to solve each system of equations.

1. $2x + y = 8$
$5x - 3y = -12$

2. $4x - 3y = 18$
$x + 2y = 12$

3. $7x - 2y = 15$
$3x + y = -10$

4. $4x - 6y = 20$
$3x + y + 8 = 0$

5. $5x + 2y = 18$
$x = -4y + 25$

6. $3x - y = 24$
$3y = 80 - 2x$

5-1 Study Guide and Intervention

Graphing Quadratic Functions

Graph Quadratic Functions

Quadratic Function	A function defined by an equation of the form $f(x) = ax^2 + bx + c$, where $a \neq 0$
Graph of a Quadratic Function	A **parabola** with these characteristics: y-intercept: c; axis of symmetry: $x = \dfrac{-b}{2a}$; x-coordinate of vertex: $\dfrac{-b}{2a}$

Example Find the y-intercept, the equation of the axis of symmetry, and the x-coordinate of the vertex for the graph of $f(x) = x^2 - 3x + 5$. Use this information to graph the function.

$a = 1$, $b = -3$, and $c = 5$, so the y-intercept is 5. The equation of the axis of symmetry is $x = \dfrac{-(-3)}{2(1)}$ or $\dfrac{3}{2}$. The x-coordinate of the vertex is $\dfrac{3}{2}$.

Next make a table of values for x near $\dfrac{3}{2}$.

x	$x^2 - 3x + 5$	$f(x)$	$(x, f(x))$
0	$0^2 - 3(0) + 5$	5	$(0, 5)$
1	$1^2 - 3(1) + 5$	3	$(1, 3)$
$\dfrac{3}{2}$	$\left(\dfrac{3}{2}\right)^2 - 3\left(\dfrac{3}{2}\right) + 5$	$\dfrac{11}{4}$	$\left(\dfrac{3}{2}, \dfrac{11}{4}\right)$
2	$2^2 - 3(2) + 5$	3	$(2, 3)$
3	$3^2 - 3(3) + 5$	5	$(3, 5)$

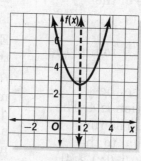

Exercises

Complete parts a–c for each quadratic function.

a. Find the y-intercept, the equation of the axis of symmetry, and the x-coordinate of the vertex.

b. Make a table of values that includes the vertex.

c. Use this information to graph the function.

1. $f(x) = x^2 + 6x + 8$

2. $f(x) = -x^2 - 2x + 2$

3. $f(x) = 2x^2 - 4x + 3$

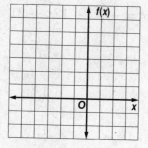

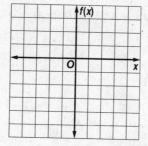

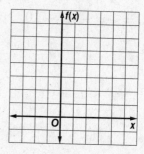

5-1 Study Guide and Intervention *(continued)*

Graphing Quadratic Functions

Maximum and Minimum Values The y-coordinate of the vertex of a quadratic function is the **maximum value** or **minimum value** of the function.

Maximum or Minimum Value of a Quadratic Function	The graph of $f(x) = ax^2 + bx + c$, where $a \neq 0$, opens up and has a minimum when $a > 0$. The graph opens down and has a maximum when $a < 0$.

Example Determine whether each function has a *maximum* or *minimum* value, and find that value. Then state the domain and range of the function.

a. $f(x) = 3x^2 - 6x + 7$

For this function, $a = 3$ and $b = -6$.

Since $a > 0$, the graph opens up, and the function has a minimum value.

The minimum value is the y-coordinate of the vertex. The x-coordinate of the vertex is $\frac{-b}{2a} = -\left(\frac{-6}{2(3)}\right) = 1$.

Evaluate the function at $x = 1$ to find the minimum value.

$f(1) = 3(1)^2 - 6(1) + 7 = 4$, so the minimum value of the function is 4. The domain is all real numbers. The range is all reals greater than or equal to the minimum value, that is $\{f(x) \mid f(x) \geq 4\}$.

b. $f(x) = 100 - 2x - x^2$

For this function, $a = -1$ and $b = -2$.

Since $a < 0$, the graph opens down, and the function has a maximum value.

The maximum value is the y-coordinate of the vertex. The x-coordinate of the vertex is $\frac{-b}{2a} = -\frac{-2}{2(-1)} = -1$.

Evaluate the function at $x = -1$ to find the maximum value.

$f(-1) = 100 - 2(-1) - (-1)^2 = 101$, so the maximum value of the function is 101. The domain is all real numbers. The range is all reals less than or equal to the maximum value, that is $\{f(x) \mid f(x) \leq 101\}$.

Exercises

Determine whether each function has a *maximum* or *minimum* value, and find that value. Then state the domain and range of the function.

1. $f(x) = 2x^2 - x + 10$

2. $f(x) = x^2 + 4x - 7$

3. $f(x) = 3x^2 - 3x + 1$

4. $f(x) = x^2 + 5x + 2$

5. $f(x) = 20 + 6x - x^2$

6. $f(x) = 4x^2 + x + 3$

7. $f(x) = -x^2 - 4x + 10$

8. $f(x) = x^2 - 10x + 5$

9. $f(x) = -6x^2 + 12x + 21$

5-2 Study Guide and Intervention

Solving Quadratic Equations by Graphing

Solve Quadratic Equations

Quadratic Equation	A quadratic equation has the form $ax^2 + bx + c = 0$, where $a \neq 0$.
Roots of a Quadratic Equation	solution(s) of the equation, or the zero(s) of the related quadratic function

The zeros of a quadratic function are the x-intercepts of its graph. Therefore, finding the x-intercepts is one way of solving the related quadratic equation.

Example Solve $x^2 + x - 6 = 0$ by graphing.

Graph the related function $f(x) = x^2 + x - 6$.

The x-coordinate of the vertex is $\frac{-b}{2a} = -\frac{1}{2}$, and the equation of the axis of symmetry is $x = -\frac{1}{2}$.

Make a table of values using x-values around $-\frac{1}{2}$.

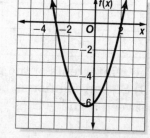

x	-1	$-\frac{1}{2}$	0	1	2
$f(x)$	-6	$-6\frac{1}{4}$	-6	-4	0

From the table and the graph, we can see that the zeros of the function are 2 and -3.

Exercises

Use the related graph of each equation to determine its solution.

1. $x^2 + 2x - 8 = 0$

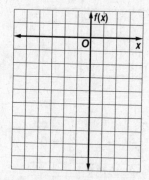

2. $x^2 - 4x - 5 = 0$

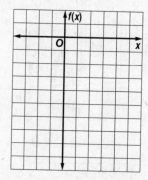

3. $x^2 - 5x + 4 = 0$

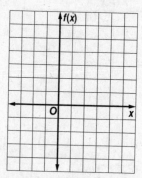

4. $x^2 - 10x + 21 = 0$

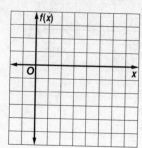

5. $x^2 + 4x + 6 = 0$

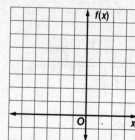

6. $4x^2 + 4x + 1 = 0$

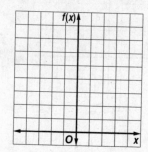

5-2 Study Guide and Intervention *(continued)*

Solving Quadratic Equations by Graphing

Estimate Solutions Often, you may not be able to find exact solutions to quadratic equations by graphing. But you can use the graph to estimate solutions.

Example Solve $x^2 - 2x - 2 = 0$ by graphing. If exact roots cannot be found, state the consecutive integers between which the roots are located.

The equation of the axis of symmetry of the related function is

$x = -\dfrac{-2}{2(1)} = 1$, so the vertex has x-coordinate 1. Make a table of values.

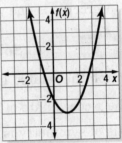

x	−1	0	1	2	3
f(x)	1	−2	−3	−2	1

The x-intercepts of the graph are between 2 and 3 and between 0 and −1. So one solution is between 2 and 3, and the other solution is between 0 and −1.

Exercises

Solve the equations. If exact roots cannot be found, state the consecutive integers between which the roots are located.

1. $x^2 - 4x + 2 = 0$

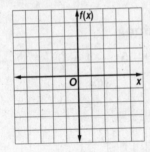

2. $x^2 + 6x + 6 = 0$

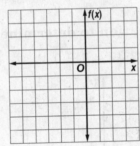

3. $x^2 + 4x + 2 = 0$

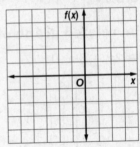

4. $-x^2 + 2x + 4 = 0$

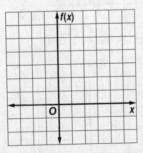

5. $2x^2 - 12x + 17 = 0$

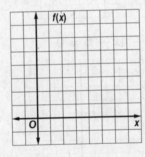

6. $-\dfrac{1}{2}x^2 + x + \dfrac{5}{2} = 0$

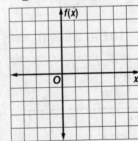

5-3 Study Guide and Intervention

Solving Quadratic Equations by Factoring

Factored Form To write a quadratic equation with roots p and q, let $(x - p)(x - q) = 0$. Then multiply using FOIL.

Example Write a quadratic equation in standard form with the given roots.

a. 3, −5

$(x - p)(x - q) = 0$	Write the pattern.
$(x - 3)[x - (-5)] = 0$	Replace p with 3, q with −5.
$(x - 3)(x + 5) = 0$	Simplify.
$x^2 + 2x - 15 = 0$	Use FOIL.

The equation $x^2 + 2x - 15 = 0$ has roots 3 and −5.

b. $-\dfrac{7}{8}, \dfrac{1}{3}$

$$(x - p)(x - q) = 0$$

$$\left[x - \left(-\frac{7}{8}\right)\right]\left(x - \frac{1}{3}\right) = 0$$

$$\left(x + \frac{7}{8}\right)\left(x - \frac{1}{3}\right) = 0$$

$$\frac{(8x + 7)}{8} \cdot \frac{(3x - 1)}{3} = 0$$

$$\frac{24 \cdot (8x + 7)(3x - 1)}{24} = 24 \cdot 0$$

$$24x^2 + 13x - 7 = 0$$

The equation $24x^2 + 13x - 7 = 0$ has roots $-\dfrac{7}{8}$ and $\dfrac{1}{3}$.

Exercises

Write a quadratic equation in standard form with the given root(s).

1. 3, −4

2. −8, −2

3. 1, 9

4. −5

5. 10, 7

6. −2, 15

7. $-\dfrac{1}{3}, 5$

8. $2, \dfrac{2}{3}$

9. $-7, \dfrac{3}{4}$

10. $3, \dfrac{2}{5}$

11. $-\dfrac{4}{9}, -1$

12. $9, \dfrac{1}{6}$

13. $\dfrac{2}{3}, -\dfrac{2}{3}$

14. $\dfrac{5}{4}, -\dfrac{1}{2}$

15. $\dfrac{3}{7}, \dfrac{1}{5}$

16. $-\dfrac{7}{8}, \dfrac{7}{2}$

17. $\dfrac{1}{2}, \dfrac{3}{4}$

18. $\dfrac{1}{8}, \dfrac{1}{6}$

5-3 Study Guide and Intervention (continued)

Solving Quadratic Equations by Factoring

Solve Equations by Factoring When you use factoring to solve a quadratic equation, you use the following property.

Zero Product Property	For any real numbers a and b, if $ab = 0$, then either $a = 0$ or $b = 0$, or both a and $b = 0$.

Example Solve each equation by factoring.

a. $3x^2 = 15x$

$3x^2 = 15x$ Original equation

$3x^2 - 15x = 0$ Subtract 15x from both sides.

$3x(x - 5) = 0$ Factor the binomial.

$3x = 0$ or $x - 5 = 0$ Zero Product Property

$x = 0$ or $x = 5$ Solve each equation.

The solution set is $\{0, 5\}$.

b. $4x^2 - 5x = 21$

$4x^2 - 5x = 21$ Original equation

$4x^2 - 5x - 21 = 0$ Subtract 21 from both sides.

$(4x + 7)(x - 3) = 0$ Factor the trinomial.

$4x + 7 = 0$ or $x - 3 = 0$ Zero Product Property

$x = -\dfrac{7}{4}$ or $x = 3$ Solve each equation.

The solution set is $\left\{-\dfrac{7}{4}, 3\right\}$.

Exercises

Solve each equation by factoring.

1. $6x^2 - 2x = 0$

2. $x^2 = 7x$

3. $20x^2 = -25x$

4. $6x^2 = 7x$

5. $6x^2 - 27x = 0$

6. $12x^2 - 8x = 0$

7. $x^2 + x - 30 = 0$

8. $2x^2 - x - 3 = 0$

9. $x^2 + 14x + 33 = 0$

10. $4x^2 + 27x - 7 = 0$

11. $3x^2 + 29x - 10 = 0$

12. $6x^2 - 5x - 4 = 0$

13. $12x^2 - 8x + 1 = 0$

14. $5x^2 + 28x - 12 = 0$

15. $2x^2 - 250x + 5000 = 0$

16. $2x^2 - 11x - 40 = 0$

17. $2x^2 + 21x - 11 = 0$

18. $3x^2 + 2x - 21 = 0$

19. $8x^2 - 14x + 3 = 0$

20. $6x^2 + 11x - 2 = 0$

21. $5x^2 + 17x - 12 = 0$

22. $12x^2 + 25x + 12 = 0$

23. $12x^2 + 18x + 6 = 0$

24. $7x^2 - 36x + 5 = 0$

5-4 Study Guide and Intervention

Complex Numbers

Pure Imaginary Numbers A **square root** of a number n is a number whose square is n. For nonnegative real numbers a and b, $\sqrt{ab} = \sqrt{a} \cdot \sqrt{b}$ and $\sqrt{\dfrac{a}{b}} = \dfrac{\sqrt{a}}{\sqrt{b}}$, $b \neq 0$.

- The **imaginary unit** i is defined to have the property that $i^2 = -1$.
- Simplified square root expressions do not have radicals in the denominator, and any number remaining under the square root has no perfect square factor other than 1.

Example 1

a. Simplify $\sqrt{-48}$.

$$\begin{aligned}
\sqrt{-48} &= \sqrt{16 \cdot (-3)} \\
&= \sqrt{16} \cdot \sqrt{3} \cdot \sqrt{-1} \\
&= 4i\sqrt{3}
\end{aligned}$$

b. Simplify $\sqrt{-63}$.

$$\begin{aligned}
\sqrt{-63} &= \sqrt{-1 \cdot 7 \cdot 9} \\
&= \sqrt{-1} \cdot \sqrt{7} \cdot \sqrt{9} \\
&= 3i\sqrt{7}
\end{aligned}$$

Example 2

a. Simplify $-3i \cdot 4i$.

$$\begin{aligned}
-3i \cdot 4i &= -12i^2 \\
&= -12(-1) \\
&= 12
\end{aligned}$$

b. Simplify $\sqrt{-3} \cdot \sqrt{-15}$.

$$\begin{aligned}
\sqrt{-3} \cdot \sqrt{-15} &= i\sqrt{3} \cdot i\sqrt{15} \\
&= i^2\sqrt{45} \\
&= -1 \cdot \sqrt{9} \cdot \sqrt{5} \\
&= -3\sqrt{5}
\end{aligned}$$

Example 3   Solve $x^2 + 5 = 0$.

$$\begin{aligned}
x^2 + 5 &= 0 && \text{Original equation.} \\
x^2 &= -5 && \text{Subtract 5 from each side.} \\
x &= \pm\sqrt{5}i && \text{Square Root Property.}
\end{aligned}$$

Exercises

Simplify.

1. $\sqrt{-72}$

2. $\sqrt{-24}$

3. $\sqrt{-84}$

4. $(2 + i)(2 - i)$

Solve each equation.

5. $5x^2 + 45 = 0$

6. $4x^2 + 24 = 0$

7. $-9x^2 = 9$

8. $7x^2 + 84 = 0$

5-4 Study Guide and Intervention *(continued)*

Complex Numbers

Operations with Complex Numbers

Complex Number	A complex number is any number that can be written in the form $a + bi$, where a and b are real numbers and i is the imaginary unit ($i^2 = -1$). a is called the real part, and b is called the imaginary part.
Addition and Subtraction of Complex Numbers	Combine like terms. $(a + bi) + (c + di) = (a + c) + (b + d)i$ $(a + bi) - (c + di) = (a - c) + (b - d)i$
Multiplication of Complex Numbers	Use the definition of i^2 and the FOIL method: $(a + bi)(c + di) = (ac - bd) + (ad + bc)i$
Complex Conjugate	$a + bi$ and $a - bi$ are complex conjugates. The product of complex conjugates is always a real number.

To divide by a complex number, first multiply the dividend and divisor by the **complex conjugate** of the divisor.

Example 1 Simplify $(6 + i) + (4 - 5i)$.

$(6 + i) + (4 - 5i)$
$= (6 + 4) + (1 - 5)i$
$= 10 - 4i$

Example 2 Simplify $(8 + 3i) - (6 - 2i)$.

$(8 + 3i) - (6 - 2i)$
$= (8 - 6) + [3 - (-2)]i$
$= 2 + 5i$

Example 3 Simplify $(2 - 5i) \cdot (-4 + 2i)$.

$(2 - 5i) \cdot (-4 + 2i)$
$= 2(-4) + 2(2i) + (-5i)(-4) + (-5i)(2i)$
$= -8 + 4i + 20i - 10i^2$
$= -8 + 24i - 10(-1)$
$= 2 + 24i$

Example 4 Simplify $\dfrac{3 - i}{2 + 3i}$.

$\dfrac{3 - i}{2 + 3i} = \dfrac{3 - i}{2 + 3i} \cdot \dfrac{2 - 3i}{2 - 3i}$
$= \dfrac{6 - 9i - 2i + 3i^2}{4 - 9i^2}$
$= \dfrac{3 - 11i}{13}$
$= \dfrac{3}{13} - \dfrac{11}{13}i$

Exercises

Simplify.

1. $(-4 + 2i) + (6 - 3i)$

2. $(5 - i) - (3 - 2i)$

3. $(6 - 3i) + (4 - 2i)$

4. $(-11 + 4i) - (1 - 5i)$

5. $(8 + 4i) + (8 - 4i)$

6. $(5 + 2i) - (-6 - 3i)$

7. $(2 + i)(3 - i)$

8. $(5 - 2i)(4 - i)$

9. $(4 - 2i)(1 - 2i)$

10. $\dfrac{5}{3 + i}$

11. $\dfrac{7 - 13i}{2i}$

12. $\dfrac{6 - 5i}{3i}$

5-5 Study Guide and Intervention

Completing the Square

Square Root Property Use the Square Root Property to solve a quadratic equation that is in the form "perfect square trinomial = constant."

Example Solve each equation by using the Square Root Property. Round to the nearest hundredth if necessary.

a. $x^2 - 8x + 16 = 25$

$x^2 - 8x + 16 = 25$

$(x - 4)^2 = 25$

$x - 4 = \sqrt{25}$ or $x - 4 = -\sqrt{25}$

$x = 5 + 4 = 9$ or $x = -5 + 4 = -1$

The solution set is $\{9, -1\}$.

b. $4x^2 - 20x + 25 = 32$

$4x^2 - 20x + 25 = 32$

$(2x - 5)^2 = 32$

$2x - 5 = \sqrt{32}$ or $2x - 5 = -\sqrt{32}$

$2x - 5 = 4\sqrt{2}$ or $2x - 5 = -4\sqrt{2}$

$x = \dfrac{5 \pm 4\sqrt{2}}{2}$

The solution set is $\left\{ \dfrac{5 \pm 4\sqrt{2}}{2} \right\}$.

Exercises

Solve each equation by using the Square Root Property. Round to the nearest hundredth if necessary.

1. $x^2 - 18x + 81 = 49$

2. $x^2 + 20x + 100 = 64$

3. $4x^2 + 4x + 1 = 16$

4. $36x^2 + 12x + 1 = 18$

5. $9x^2 - 12x + 4 = 4$

6. $25x^2 + 40x + 16 = 28$

7. $4x^2 - 28x + 49 = 64$

8. $16x^2 + 24x + 9 = 81$

9. $100x^2 - 60x + 9 = 121$

10. $25x^2 + 20x + 4 = 75$

11. $36x^2 + 48x + 16 = 12$

12. $25x^2 - 30x + 9 = 96$

5-5 Study Guide and Intervention *(continued)*

Completing the Square

Complete the Square To complete the square for a quadratic expression of the form $x^2 + bx$, follow these steps.

1. Find $\dfrac{b}{2}$. → 2. Square $\dfrac{b}{2}$. → 3. Add $\left(\dfrac{b}{2}\right)^2$ to $x^2 + bx$.

Example 1 Find the value of c that makes $x^2 + 22x + c$ a perfect square trinomial. Then write the trinomial as the square of a binomial.

Step 1 $b = 22$; $\dfrac{b}{2} = 11$

Step 2 $11^2 = 121$

Step 3 $c = 121$

The trinomial is $x^2 + 22x + 121$, which can be written as $(x + 11)^2$.

Example 2 Solve $2x^2 - 8x - 24 = 0$ by completing the square.

$2x^2 - 8x - 24 = 0$	Original equation
$\dfrac{2x^2 - 8x - 24}{2} = \dfrac{0}{2}$	Divide each side by 2.
$x^2 - 4x - 12 = 0$	$x^2 - 4x - 12$ is not a perfect square.
$x^2 - 4x = 12$	Add 12 to each side.
$x^2 - 4x + 4 = 12 + 4$	Since $\left(\dfrac{4}{2}\right)^2 = 4$, add 4 to each side.
$(x - 2)^2 = 16$	Factor the square.
$x - 2 = \pm 4$	Square Root Property
$x = 6$ or $x = -2$	Solve each equation.

The solution set is $\{6, -2\}$.

Exercises

Find the value of c that makes each trinomial a perfect square. Then write the trinomial as a perfect square.

1. $x^2 - 10x + c$

2. $x^2 + 60x + c$

3. $x^2 - 3x + c$

4. $x^2 + 3.2x + c$

5. $x^2 + \dfrac{1}{2}x + c$

6. $x^2 - 2.5x + c$

Solve each equation by completing the square.

7. $y^2 - 4y - 5 = 0$

8. $x^2 - 8x - 65 = 0$

9. $w^2 - 10w + 21 = 0$

10. $2x^2 - 3x + 1 = 0$

11. $2x^2 - 13x - 7 = 0$

12. $25x^2 + 40x - 9 = 0$

13. $x^2 + 4x + 1 = 0$

14. $y^2 + 12y + 4 = 0$

15. $t^2 + 3t - 8 = 0$

5-6 Study Guide and Intervention

The Quadratic Formula and the Discriminant

Quadratic Formula The **Quadratic Formula** can be used to solve *any* quadratic equation once it is written in the form $ax^2 + bx + c = 0$.

Quadratic Formula	The solutions of $ax^2 + bx + c = 0$, with $a \neq 0$, are given by $x = \dfrac{-b \pm \sqrt{b^2 - 4ac}}{2a}$.

Example Solve $x^2 - 5x = 14$ by using the Quadratic Formula.

Rewrite the equation as $x^2 - 5x - 14 = 0$.

$$x = \frac{-b \pm \sqrt{b^2 - 4ac}}{2a} \qquad \text{Quadratic Formula}$$

$$= \frac{-(-5) \pm \sqrt{(-5)^2 - 4(1)(-14)}}{2(1)} \qquad \text{Replace } a \text{ with 1, } b \text{ with } -5, \text{ and } c \text{ with } -14.$$

$$= \frac{5 \pm \sqrt{81}}{2} \qquad \text{Simplify.}$$

$$= \frac{5 \pm 9}{2}$$

$$= 7 \text{ or } -2$$

The solutions are -2 and 7.

Exercises

Solve each equation by using the Quadratic Formula.

1. $x^2 + 2x - 35 = 0$

2. $x^2 + 10x + 24 = 0$

3. $x^2 - 11x + 24 = 0$

4. $4x^2 + 19x - 5 = 0$

5. $14x^2 + 9x + 1 = 0$

6. $2x^2 - x - 15 = 0$

7. $3x^2 + 5x = 2$

8. $2y^2 + y - 15 = 0$

9. $3x^2 - 16x + 16 = 0$

10. $8x^2 + 6x - 9 = 0$

11. $r^2 - \dfrac{3r}{5} + \dfrac{2}{25} = 0$

12. $x^2 - 10x - 50 = 0$

13. $x^2 + 6x - 23 = 0$

14. $4x^2 - 12x - 63 = 0$

15. $x^2 - 6x + 21 = 0$

5-6 Study Guide and Intervention *(continued)*

The Quadratic Formula and the Discriminant

Roots and the Discriminant

Discriminant	The expression under the radical sign, $b^2 - 4ac$, in the Quadratic Formula is called the **discriminant**.

Discriminant	Type and Number of Roots
$b^2 - 4ac > 0$ and a perfect square	2 rational roots
$b^2 - 4ac > 0$, but **not** a perfect square	2 irrational roots
$b^2 - 4ac = 0$	1 rational root
$b^2 - 4ac < 0$	2 complex roots

Example Find the value of the discriminant for each equation. Then describe the number and type of roots for the equation.

a. $2x^2 + 5x + 3$
The discriminant is
$b^2 - 4ac = 5^2 - 4(2)(3)$ or 1.
The discriminant is a perfect square, so the equation has 2 rational roots.

b. $3x^2 - 2x + 5$
The discriminant is
$b^2 - 4ac = (-2)^2 - 4(3)(5)$ or -56.
The discriminant is negative, so the equation has 2 complex roots.

Exercises

Complete parts a–c for each quadratic equation.

a. Find the value of the discriminant.
b. Describe the number and type of roots.
c. Find the exact solutions by using the Quadratic Formula.

1. $p^2 + 12p = -4$

2. $9x^2 - 6x + 1 = 0$

3. $2x^2 - 7x - 4 = 0$

4. $x^2 + 4x - 4 = 0$

5. $5x^2 - 36x + 7 = 0$

6. $4x^2 - 4x + 11 = 0$

7. $x^2 - 7x + 6 = 0$

8. $m^2 - 8m = -14$

9. $25x^2 - 40x = -16$

10. $4x^2 + 20x + 29 = 0$

11. $6x^2 + 26x + 8 = 0$

12. $4x^2 - 4x - 11 = 0$

5-7 **Study Guide and Intervention**

Transformations with Quadratic Functions

Write Quadratic Equations in Vertex Form A quadratic function is easier to graph when it is in vertex form. You can write a quadratic function of the form $y = ax^2 + bx + c$ in vertex from by completing the square.

Example Write $y = 2x^2 - 12x + 25$ in vertex form. Then graph the function.

$y = 2x^2 - 12x + 25$

$y = 2(x^2 - 6x) + 25$

$y = 2(x^2 - 6x + 9) + 25 - 18$

$y = 2(x - 3)^2 + 7$

The vertex form of the equation is $y = 2(x - 3)^2 + 7$.

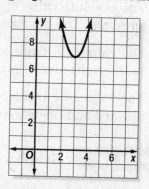

Exercises

Write each equation in vertex form. Then graph the function.

1. $y = x^2 - 10x + 32$ **2.** $y = x^2 + 6x$ **3.** $y = x^2 - 8x + 6$

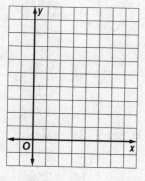   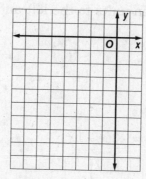   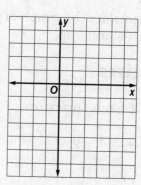

4. $y = -4x^2 + 16x - 11$ **5.** $y = 3x^2 - 12x + 5$ **6.** $y = 5x^2 - 10x + 9$

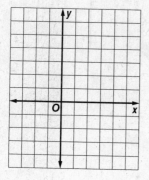   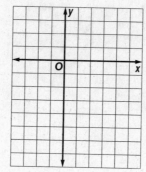   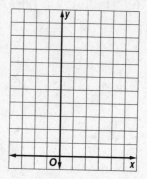

5-7 Study Guide and Intervention (continued)

Transformations with Quadratic Functions

Transformations of Quadratic Functions Parabolas can be transformed by changing the values of the constants a, h, and k in the vertex form of a quadratic equation: $y = a(x - h) + k$.

- The sign of a determines whether the graph opens upward ($a > 0$) or downward ($a < 0$).

- The absolute value of a also causes a dilation (enlargement or reduction) of the parabola. The parabola becomes narrower if $|a| > 1$ and wider if $|a| < 1$.

- The value of h translates the parabola horizontally. Positive values of h slide the graph to the right and negative values slide the graph to the left.

- The value of k translates the graph vertically. Positive values of k slide the graph upward and negative values slide the graph downward.

| **Example** | **Graph $y = (x + 7)^2 + 3$.** |

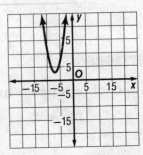

- Rewrite the equation as $y = [x - (-7)]^2 + 3$.

- Because $h = -7$ and $k = 3$, the vertex is at $(-7, 3)$. The axis of symmetry is $x = -7$. Because $a = 1$, we know that the graph opens up, and the graph is the same width as the graph of $y = x^2$.

- Translate the graph of $y = x^2$ seven units to the left and three units up.

Exercises

Graph each function.

1. $y = -2x^2 + 2$

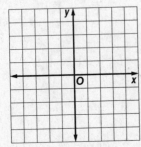

2. $y = -3(x - 1)^2$

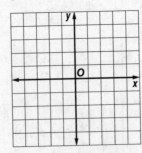

3. $y = 2(x + 2)^2 + 3$

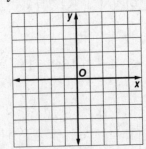

5-8 Study Guide and Intervention

Quadratic Inequalities

Graph Quadratic Inequalities To graph a quadratic inequality in two variables, use the following steps:

1. Graph the related quadratic equation, $y = ax^2 + bx + c$.
   Use a dashed line for $<$ or $>$; use a solid line for $\leq$ or $\geq$.

2. Test a point inside the parabola.
   If it satisfies the inequality, shade the region inside the parabola; otherwise, shade the region outside the parabola.

Example **Graph the inequality $y > x^2 + 6x + 7$.**

First graph the equation $y = x^2 + 6x + 7$. By completing the square, you get the vertex form of the equation $y = (x + 3)^2 - 2$, so the vertex is $(-3, -2)$. Make a table of values around $x = -3$, and graph. Since the inequality includes $>$, use a dashed line.

Test the point $(-3, 0)$, which is inside the parabola. Since $(-3)^2 + 6(-3) + 7 = -2$, and $0 > -2$, $(-3, 0)$ satisfies the inequality. Therefore, shade the region inside the parabola.

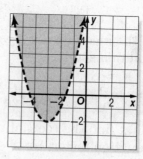

Exercises

Graph each inequality.

1. $y > x^2 - 8x + 17$

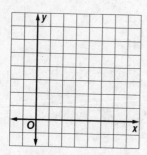

2. $y \leq x^2 + 6x + 4$

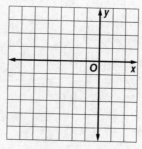

3. $y \geq x^2 + 2x + 2$

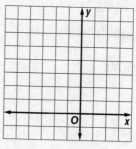

4. $y < -x^2 + 4x - 6$

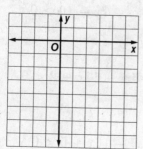

5. $y \geq 2x^2 + 4x$

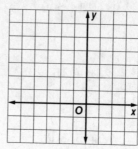

6. $y > -2x^2 - 4x + 2$

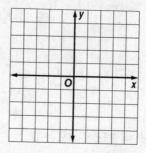

5-8 Study Guide and Intervention *(continued)*

Quadratic Inequalities

Solve Quadratic Inequalities Quadratic inequalities in one variable can be solved graphically or algebraically.

Graphical Method	To solve $ax^2 + bx + c < 0$: First graph $y = ax^2 + bx + c$. The solution consists of the x-values for which the graph is **below** the x-axis. To solve $ax^2 + bx + c > 0$: First graph $y = ax^2 + bx + c$. The solution consists of the x-values for which the graph is **above** the x-axis.
Algebraic Method	Find the roots of the related quadratic equation by factoring, completing the square; or using the Quadratic Formula. 2 roots divide the number line into 3 intervals. Test a value in each interval to see which intervals are solutions.

If the inequality involves ≤ or ≥, the roots of the related equation are included in the solution set.

Example **Solve the inequality $x^2 - x - 6 \le 0$.**

First find the roots of the related equation $x^2 - x - 6 = 0$. The equation factors as $(x - 3)(x + 2) = 0$, so the roots are 3 and -2. The graph opens up with x-intercepts 3 and -2, so it must be on or below the x-axis for $-2 \le x \le 3$. Therefore the solution set is $\{x|-2 \le x \le 3\}$.

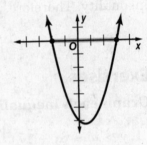

Exercises

Solve each inequality.

1. $x^2 + 2x < 0$

2. $x^2 - 16 < 0$

3. $0 < 6x - x^2 - 5$

4. $c^2 \le 4$

5. $2m^2 - m < 1$

6. $y^2 < -8$

7. $x^2 - 4x - 12 < 0$

8. $x^2 + 9x + 14 > 0$

9. $-x^2 + 7x - 10 \ge 0$

10. $2x^2 + 5x - 3 \le 0$

11. $4x^2 - 23x + 15 > 0$

12. $-6x^2 - 11x + 2 < 0$

13. $2x^2 - 11x + 12 \ge 0$

14. $x^2 - 4x + 5 < 0$

15. $3x^2 - 16x + 5 < 0$

6-1 Study Guide and Intervention

Operations with Polynomials

Multiply and Divide Monomials **Negative exponents** are a way of expressing the multiplicative inverse of a number.

Negative Exponents	$a^{-n} = \dfrac{1}{a^n}$ and $\dfrac{1}{a^{-n}} = a^n$ for any real number $a \neq 0$ and any integer n.

When you **simplify an expression**, you rewrite it without powers of powers, parentheses, or negative exponents. Each base appears only once, and all fractions are in simplest form. The following properties are useful when simplifying expressions.

Product of Powers	$a^m \cdot a^n = a^{m+n}$ for any real number a and integers m and n.
Quotient of Powers	$\dfrac{a^m}{a^n} = a^{m-n}$ for any real number $a \neq 0$ and integers m and n.
Properties of Powers	For a, b real numbers and m, n integers: $(a^m)^n = a^{mn}$ $(ab)^m = a^m b^m$ $\left(\dfrac{a}{b}\right)^n = \dfrac{a^n}{b^n}, b \neq 0$ $\left(\dfrac{a}{b}\right)^{-n} = \left(\dfrac{b}{a}\right)^n$ or $\dfrac{b^n}{a^n}, a \neq 0, b \neq 0$

Example Simplify. Assume that no variable equals 0.

a. $(3m^4 n^{-2})(-5mn)^2$

$(3m^4 n^{-2})(-5mn)^2 = 3m^4 n^{-2} \cdot 25m^2 n^2$
$= 75m^4 m^2 n^{-2} n^2$
$= 75m^{4+2} n^{-2+2}$
$= 75m^6$

b. $\dfrac{(-m^4)^3}{(2m^2)^{-2}}$

$\dfrac{(-m^4)^3}{(2m^2)^{-2}} = \dfrac{-m^{12}}{\dfrac{1}{4m^4}}$
$= -m^{12} \cdot 4m^4$
$= -4m^{16}$

Exercises

Simplify. Assume that no variable equals 0.

1. $c^{12} \cdot c^{-4} \cdot c^6$

2. $\dfrac{b^8}{b^2}$

3. $(a^4)^5$

4. $\dfrac{x^{-2}y}{x^4 y^{-1}}$

5. $\left(\dfrac{a^2 b}{a^{-3} b^2}\right)^{-1}$

6. $\left(\dfrac{x^2 y}{xy^3}\right)^2$

7. $\dfrac{1}{2}(-5a^2 b^3)^2 (abc)^2$

8. $m^7 \cdot m^8$

9. $\dfrac{8m^3 n^2}{4mn^3}$

10. $\dfrac{2^3 c^4 t^2}{2^2 c^4 t^2}$

11. $4j(-j^{-2}k^2)(3j^3 k^{-7})$

12. $\dfrac{2mn^2(3m^2 n)^2}{12m^3 n^4}$

6-1 Study Guide and Intervention *(continued)*

Operations with Polynomials

Operations with Polynomials

Polynomial	a monomial or a sum of monomials
Like Terms	terms that have the same variable(s) raised to the same power(s)

To add or subtract polynomials, perform the indicated operations and combine like terms.

Example 1 Simplify $4xy^2 + 12xy - 7x^2y - (20xy + 5xy^2 - 8x^2y)$.

$4xy^2 + 12xy - 7x^2y - (20xy + 5xy^2 - 8x^2y)$

$= 4xy^2 + 12xy - 7x^2y - 20xy - 5xy^2 + 8x^2y$ Distribute the minus sign.

$= (-7x^2y + 8x^2y) + (4xy^2 - 5xy^2) + (12xy - 20xy)$ Group like terms.

$= x^2y - xy^2 - 8xy$ Combine like terms.

You use the distributive property when you multiply polynomials. When multiplying binomials, the **FOIL** pattern is helpful.

FOIL Pattern	To multiply two binomials, add the products of
	F the *first* terms, **O** the *outer* terms, **I** the *inner* terms, and **L** the *last* terms.

Example 2 Find $(6x - 5)(2x + 1)$.

$(6x - 5)(2x + 1) = \underset{\text{First terms}}{6x \cdot 2x} + \underset{\text{Outer terms}}{6x \cdot 1} + \underset{\text{Inner terms}}{(-5) \cdot 2x} + \underset{\text{Last terms}}{(-5) \cdot 1}$

$= 12x^2 + 6x - 10x - 5$ Multiply monomials.

$= 12x^2 - 4x - 5$ Add like terms.

Exercises

Simplify.

1. $(6x^2 - 3x + 2) - (4x^2 + x - 3)$

2. $(7y^2 + 12xy - 5x^2) + (6xy - 4y^2 - 3x^2)$

3. $(-4m^2 - 6m) - (6m + 4m^2)$

4. $27x^2 - 5y^2 + 12y^2 - 14x^2$

5. $\frac{1}{4}x^2 - \frac{3}{8}xy + \frac{1}{2}y^2 - \frac{1}{2}xy + \frac{1}{4}y^2 - \frac{3}{8}x^2$

6. $24p^3 - 15p^2 + 3p - 15p^3 + 13p^2 - 7p$

Find each product.

7. $2x(3x^2 - 5)$

8. $7a(6 - 2a - a^2)$

9. $(x^2 - 2)(x^2 - 5)$

10. $(x + 1)(2x^2 - 3x + 1)$

11. $(2n^2 - 3)(n^2 + 5n - 1)$

12. $(x - 1)(x^2 - 3x + 4)$

6-2 Study Guide and Intervention

Dividing Polynomials

Long Division To divide a polynomial by a monomial, use the skills learned in Lesson 6-1.

To divide a polynomial by a polynomial, use a long division pattern. Remember that only like terms can be added or subtracted.

Example 1 Simplify $\dfrac{12p^3t^2r - 21p^2qtr^2 - 9p^3tr}{3p^2tr}$.

$$\dfrac{12p^3t^2r - 21p^2qtr^2 - 9p^3tr}{3p^2tr} = \dfrac{12p^3t^2r}{3p^2tr} - \dfrac{21p^2qtr^2}{3p^2tr} - \dfrac{9p^3tr}{3p^2tr}$$

$$= \dfrac{12}{3}p^{(3-2)}t^{(2-1)}r^{(1-1)} - \dfrac{21}{3}p^{(2-2)}qt^{(1-1)}r^{(2-1)} - \dfrac{9}{3}p^{(3-2)}t^{(1-1)}r^{(1-1)}$$

$$= 4pt - 7qr - 3p$$

Example 2 Use long division to find $(x^3 - 8x^2 + 4x - 9) \div (x - 4)$.

$$
\begin{array}{r}
x^2 - 4x - 12 \\
x - 4\overline{)x^3 - 8x^2 + 4x - 9} \\
\underline{(-)x^3 - 4x^2} \\
-4x^2 + 4x \\
\underline{(-)-4x^2 + 16x} \\
-12x - 9 \\
\underline{(-)-12x + 48} \\
-57
\end{array}
$$

The quotient is $x^2 - 4x - 12$, and the remainder is -57.

Therefore $\dfrac{x^3 - 8x^2 + 4x - 9}{x - 4} = x^2 - 4x - 12 - \dfrac{57}{x - 4}$.

Exercises

Simplify.

1. $\dfrac{18a^3 + 30a^2}{3a}$

2. $\dfrac{24mn^6 - 40m^2n^3}{4m^2n^3}$

3. $\dfrac{60a^2b^3 - 48b^4 + 84a^5b^2}{12ab^2}$

4. $(2x^2 - 5x - 3) \div (x - 3)$

5. $(m^2 - 3m - 7) \div (m + 2)$

6. $(p^3 - 6) \div (p - 1)$

7. $(t^3 - 6t^2 + 1) \div (t + 2)$

8. $(x^5 - 1) \div (x - 1)$

9. $(2x^3 - 5x^2 + 4x - 4) \div (x - 2)$

6-2 Study Guide and Intervention *(continued)*

Dividing Polynomials

Synthetic Division

Synthetic division	a procedure to divide a polynomial by a binomial using coefficients of the dividend and the value of r in the divisor $x - r$

Use synthetic division to find $(2x^3 - 5x^2 + 5x - 2) \div (x - 1)$.

Step 1	Write the terms of the dividend so that the degrees of the terms are in descending order. Then write just the coefficients.	$2x^3 - 5x^2 + 5x - 2$ 2 −5 5 −2
Step 2	Write the constant r of the divisor $x - r$ to the left, In this case, $r = 1$. Bring down the first coefficient, 2, as shown.	1⌋ 2 −5 5 −2 ————————— 2
Step 3	Multiply the first coefficient by r, $1 \cdot 2 = 2$. Write their product under the second coefficient. Then add the product and the second coefficient: $-5 + 2 = -3$.	1⌋ 2 −5 5 −2 2 ————————— 2 −3
Step 4	Multiply the sum, −3, by r: $-3 \cdot 1 = -3$. Write the product under the next coefficient and add: $5 + (-3) = 2$.	1⌋ 2 −5 5 −2 2 −3 ————————— 2 −3 2
Step 5	Multiply the sum, 2, by r: $2 \cdot 1 = 2$. Write the product under the next coefficient and add: $-2 + 2 = 0$. The remainder is 0.	1⌋ 2 −5 5 −2 2 −3 2 ————————— 2 −3 2 0

Thus, $(2x^3 - 5x^2 + 5x - 2) \div (x - 1) = 2x^2 - 3x + 2$.

Exercises

Simplify.

1. $(3x^3 - 7x^2 + 9x - 14) \div (x - 2)$

2. $(5x^3 + 7x^2 - x - 3) \div (x + 1)$

3. $(2x^3 + 3x^2 - 10x - 3) \div (x + 3)$

4. $(x^3 - 8x^2 + 19x - 9) \div (x - 4)$

5. $(2x^3 + 10x^2 + 9x + 38) \div (x + 5)$

6. $(3x^3 - 8x^2 + 16x - 1) \div (x - 1)$

7. $(x^3 - 9x^2 + 17x - 1) \div (x - 2)$

8. $(4x^3 - 25x^2 + 4x + 20) \div (x - 6)$

9. $(6x^3 + 28x^2 - 7x + 9) \div (x + 5)$

10. $(x^4 - 4x^3 + x^2 + 7x - 2) \div (x - 2)$

11. $(12x^4 + 20x^3 - 24x^2 + 20x + 35) \div (3x + 5)$

6-3 Study Guide and Intervention

Polynomial Functions

Polynomial Functions

Polynomial in One Variable	A polynomial of degree n in one variable x is an expression of the form $a_n x^n + a_{n-1} x^{n-1} + \ldots + a_2 x^2 + a_1 x + a_0$, where the coefficients a_{n-1}, a_{n-2}, a_{n-3}, ..., a_0 represent real numbers, a_n is not zero, and n represents a nonnegative integer.

The **degree of a polynomial** in one variable is the greatest exponent of its variable. The **leading coefficient** is the coefficient of the term with the highest degree.

Polynomial Function	A polynomial function of degree n can be described by an equation of the form $P(x) = a_n x^n + a_{n-1} x^{n-1} + \ldots + a_2 x^2 + a_1 x + a_0$, where the coefficients a_{n-1}, a_{n-2}, a_{n-3}, ..., a_0 represent real numbers, a_n is not zero, and n represents a nonnegative integer.

Example 1 **What are the degree and leading coefficient of $3x^2 - 2x^4 - 7 + x^3$?**

Rewrite the expression so the powers of x are in decreasing order.

$-2x^4 + x^3 + 3x^2 - 7$

This is a polynomial in one variable. The degree is 4, and the leading coefficient is -2.

Example 2 **Find $f(-5)$ if $f(x) = x^3 + 2x^2 - 10x + 20$.**

$f(x) = x^3 + 2x^2 - 10x + 20$ Original function

$f(-5) = (-5)^3 + 2(-5)^2 - 10(-5) + 20$ Replace x with -5.

$\qquad = -125 + 50 + 50 + 20$ Evaluate.

$\qquad = -5$ Simplify.

Example 3 **Find $g(a^2 - 1)$ if $g(x) = x^2 + 3x - 4$.**

$g(x) = x^2 + 3x - 4$ Original function

$g(a^2 - 1) = (a^2 - 1)^2 + 3(a^2 - 1) - 4$ Replace x with $a^2 - 1$.

$\qquad = a^4 - 2a^2 + 1 + 3a^2 - 3 - 4$ Evaluate.

$\qquad = a^4 + a^2 - 6$ Simplify.

Exercises

State the degree and leading coefficient of each polynomial in one variable. If it is not a polynomial in one variable, explain why.

1. $3x^4 + 6x^3 - x^2 + 12$

2. $100 - 5x^3 + 10x^7$

3. $4x^6 + 6x^4 + 8x^8 - 10x^2 + 20$

4. $4x^2 - 3xy + 16y^2$

5. $8x^3 - 9x^5 + 4x^2 - 36$

6. $\dfrac{x^2}{18} - \dfrac{x^6}{25} + \dfrac{x^3}{36} - \dfrac{1}{72}$

Find $f(2)$ and $f(-5)$ for each function.

7. $f(x) = x^2 - 9$

8. $f(x) = 4x^3 - 3x^2 + 2x - 1$

9. $f(x) = 9x^3 - 4x^2 + 5x + 7$

6-3 Study Guide and Intervention (continued)

Polynomial Functions

Graphs of Polynomial Functions

End Behavior of Polynomial Functions	If the degree is even and the leading coefficient is positive, then
	$f(x) \to +\infty$ as $x \to -\infty$ $f(x) \to +\infty$ as $x \to +\infty$
	If the degree is even and the leading coefficient is negative, then
	$f(x) \to -\infty$ as $x \to -\infty$ $f(x) \to -\infty$ as $x \to +\infty$
	If the degree is odd and the leading coefficient is positive, then
	$f(x) \to -\infty$ as $x \to -\infty$ $f(x) \to +\infty$ as $x \to +\infty$
	If the degree is odd and the leading coefficient is negative, then
	$f(x) \to +\infty$ as $x \to -\infty$ $f(x) \to -\infty$ as $x \to +\infty$

Real Zeros of a Polynomial Function	The maximum number of zeros of a polynomial function is equal to the degree of the polynomial. A zero of a function is a point at which the graph intersects the x-axis. On a graph, count the number of real zeros of the function by counting the number of times the graph crosses or touches the x-axis.

Example Determine whether the graph represents an odd-degree polynomial or an even-degree polynomial. Then state the number of real zeros.

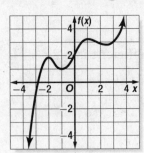

As $x \to -\infty$, $f(x) \to -\infty$ and as $x \to +\infty$, $f(x) \to +\infty$, so it is an odd-degree polynomial function.
The graph intersects the x-axis at 1 point, so the function has 1 real zero.

Exercises

For each graph,
a. describe the end behavior,
b. determine whether it represents an odd-degree or an even-degree function, and
c. state the number of real zeroes.

1.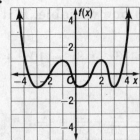

2.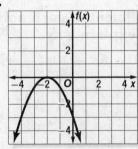

3.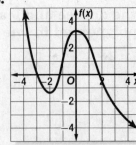

6-4 Study Guide and Intervention

Analyzing Graphs of Polynomial Functions

Graphs of Polynomial Functions

Location Principle	Suppose $y = f(x)$ represents a polynomial function and a and b are two numbers such that $f(a) < 0$ and $f(b) > 0$. Then the function has at least one real zero between a and b.

Example Determine consecutive integer values of x between which each real zero of $f(x) = 2x^4 - x^3 - 5$ is located. Then draw the graph.

Make a table of values. Look at the values of $f(x)$ to locate the zeros. Then use the points to sketch a graph of the function.

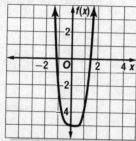

x	f(x)
-2	35
-1	-2
0	-5
1	-4
2	19

The changes in sign indicate that there are zeros between $x = -2$ and $x = -1$ and between $x = 1$ and $x = 2$.

Exercises

Graph each function by making a table of values. Determine the values of x between which each real zero is located.

1. $f(x) = x^3 - 2x^2 + 1$

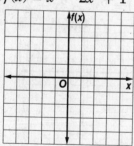

2. $f(x) = x^4 + 2x^3 - 5$

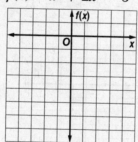

3. $f(x) = -x^4 + 2x^2 - 1$

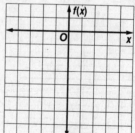

4. $f(x) = x^3 - 3x^2 + 4$

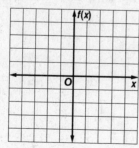

5. $f(x) = 3x^3 + 2x - 1$

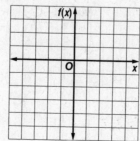

6. $f(x) = x^4 - 3x^3 + 1$

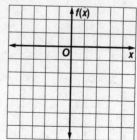

6-4 Study Guide and Intervention (continued)

Analyzing Graphs of Polynomial Functions

Maximum and Minimum Points A quadratic function has either a maximum or a minimum point on its graph. For higher degree polynomial functions, you can find *turning points*, which represent **relative maximum** or **relative minimum** points.

Example Graph $f(x) = x^3 + 6x^2 - 3$. Estimate the x-coordinates at which the relative maxima and minima occur.

Make a table of values and graph the function.

x	$f(x)$
-5	22
-4	29	← indicates a relative maximum
-3	24
-2	13
-1	2	← zero between $x = -1$, $x = 0$
0	-3	← indicates a relative minimum
1	4
2	29

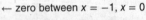
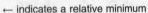

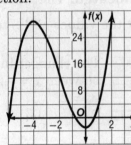

A relative maximum occurs at $x = -4$ and a relative minimum occurs at $x = 0$.

Exercises

Graph each polynomial function. Estimate the x-coordinates at which the relative maxima and relative minima occur.

1. $f(x) = x^3 - 3x^2$

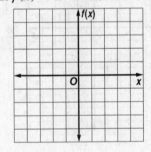

2. $f(x) = 2x^3 + x^2 - 3x$

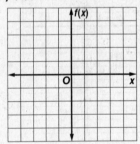

3. $f(x) = 2x^3 - 3x + 2$

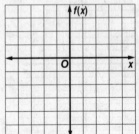

4. $f(x) = x^4 - 7x - 3$

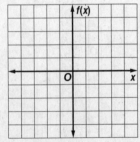

5. $f(x) = x^5 - 2x^2 + 2$

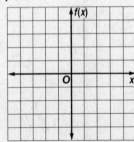

6. $f(x) = x^3 + 2x^2 - 3$

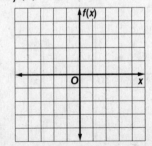

6-5 Study Guide and Intervention
Solving Polynomial Equations

Factor Polynomials

Techniques for Factoring Polynomials	For any number of terms, check for: greatest common factor
	For two terms, check for: Difference of two squares $a^2 - b^2 = (a + b)(a - b)$ Sum of two cubes $a^3 + b^3 = (a + b)(a^2 - ab + b^2)$ Difference of two cubes $a^3 - b^3 = (a - b)(a^2 + ab + b^2)$
	For three terms, check for: Perfect square trinomials $a^2 + 2ab + b^2 = (a + b)^2$ $a^2 - 2ab + b^2 = (a - b)^2$ General trinomials $acx^2 + (ad + bc)x + bd = (ax + b)(cx + d)$
	For four or more terms, check for: Grouping $ax + bx + ay + by = x(a + b) + y(a + b)$ $\qquad\qquad\qquad\quad = (a + b)(x + y)$

Example **Factor $24x^2 - 42x - 45$.**

First factor out the GCF to get $24x^2 - 42x - 45 = 3(8x^2 - 14x - 15)$. To find the coefficients of the x terms, you must find two numbers whose product is $8 \cdot (-15) = -120$ and whose sum is -14. The two coefficients must be -20 and 6. Rewrite the expression using $-20x$ and $6x$ and factor by grouping.

$$8x^2 - 14x - 15 = 8x^2 - 20x + 6x - 15 \qquad \text{Group to find a GCF.}$$
$$= 4x(2x - 5) + 3(2x - 5) \qquad \text{Factor the GCF of each binomial.}$$
$$= (4x + 3)(2x - 5) \qquad \text{Distributive Property}$$

Thus, $24x^2 - 42x - 45 = 3(4x + 3)(2x - 5)$.

Exercises

Factor completely. If the polynomial is not factorable, write *prime*.

1. $14x^2y^2 + 42xy^3$

2. $6mn + 18m - n - 3$

3. $2x^2 + 18x + 16$

4. $x^4 - 1$

5. $35x^3y^4 - 60x^4y$

6. $2r^3 + 250$

7. $100m^8 - 9$

8. $x^2 + x + 1$

9. $c^4 + c^3 - c^2 - c$

6-5   Study Guide and Intervention   *(continued)*

Solving Polynomial Equations

Solve Polynomial Equations If a polynomial expression can be written in quadratic form, then you can use what you know about solving quadratic equations to solve the related polynomial equation.

Example 1   Solve $x^4 - 40x^2 + 144 = 0$.

$x^4 - 40x^2 + 144 = 0$	Original equation
$(x^2)^2 - 40(x^2) + 144 = 0$	Write the expression on the left in quadratic form.
$(x^2 - 4)(x^2 - 36) = 0$	Factor.
$x^2 - 4 = 0$ or $x^2 - 36 = 0$	Zero Product Property
$(x - 2)(x + 2) = 0$ or $(x - 6)(x + 6) = 0$	Factor.
$x - 2 = 0$ or $x + 2 = 0$ or $x - 6 = 0$ or $x + 6 = 0$	Zero Product Property
$x = 2$ or $x = -2$ or $x = 6$ or $x = -6$	Simplify.

The solutions are ± 2 and ± 6.

Example 2   Solve $2x + \sqrt{x} - 15 = 0$.

$2x + \sqrt{x} - 15 = 0$	Original equation
$2(\sqrt{x})^2 + \sqrt{x} - 15 = 0$	Write the expression on the left in quadratic form.
$(2\sqrt{x} - 5)(\sqrt{x} + 3) = 0$	Factor.
$2\sqrt{x} - 5 = 0$ or $\sqrt{x} + 3 = 0$	Zero Product Property
$\sqrt{x} = \dfrac{5}{2}$ or $\sqrt{x} = -3$	Simplify.

Since the principal square root of a number cannot be negative, $\sqrt{x} = -3$ has no solution.
The solution is $\dfrac{25}{4}$ or $6\dfrac{1}{4}$.

Exercises

Solve each equation.

1. $x^4 = 49$

2. $x^4 - 6x^2 = -8$

3. $x^4 - 3x^2 = 54$

4. $3t^6 - 48t^2 = 0$

5. $m^6 - 16m^3 + 64 = 0$

6. $y^4 - 5y^2 + 4 = 0$

7. $x^4 - 29x^2 + 100 = 0$

8. $4x^4 - 73x^2 + 144 = 0$

9. $\dfrac{1}{x^2} - \dfrac{7}{x} + 12 = 0$

10. $x - 5\sqrt{x} + 6 = 0$

11. $x - 10\sqrt{x} + 21 = 0$

12. $x^{\frac{2}{3}} - 5x^{\frac{1}{3}} + 6 = 0$

6-6 Study Guide and Intervention
The Remainder and Factor Theorems

Synthetic Substitution

Remainder Theorem	The remainder, when you divide the polynomial $f(x)$ by $(x - a)$, is the constant $f(a)$. $f(x) = q(x) \cdot (x - a) + f(a)$, where $q(x)$ is a polynomial with degree one less than the degree of $f(x)$.

Example 1   **If $f(x) = 3x^4 + 2x^3 - 5x^2 + x - 2$, find $f(-2)$.**

Method 1 Synthetic Substitution

By the Remainder Theorem, $f(-2)$ should be the remainder when you divide the polynomial by $x + 2$.

$$\begin{array}{r|rrrrr} -2 & 3 & 2 & -5 & 1 & -2 \\ & & -6 & 8 & -6 & 10 \\ \hline & 3 & -4 & 3 & -5 & 8 \end{array}$$

The remainder is 8, so $f(-2) = 8$.

Method 2 Direct Substitution

Replace x with -2.

$f(x) = 3x^4 + 2x^3 - 5x^2 + x - 2$
$f(-2) = 3(-2)^4 + 2(-2)^3 - 5(-2)^2 + (-2) - 2$
$\quad = 48 - 16 - 20 - 2 - 2$ or 8
So $f(-2) = 8$.

Example 2   **If $f(x) = 5x^3 + 2x - 1$, find $f(3)$.**

Again, by the Remainder Theorem, $f(3)$ should be the remainder when you divide the polynomial by $x - 3$.

$$\begin{array}{r|rrrr} 3 & 5 & 0 & 2 & -1 \\ & & 15 & 45 & 141 \\ \hline & 5 & 15 & 47 & 140 \end{array}$$

The remainder is 140, so $f(3) = 140$.

Exercises

Use synthetic substitution to find $f(-5)$ and $f(\frac{1}{2})$ for each function.

1. $f(x) = -3x^2 + 5x - 1$

2. $f(x) = 4x^2 + 6x - 7$

3. $f(x) = -x^3 + 3x^2 - 5$

4. $f(x) = x^4 + 11x^2 - 1$

Use synthetic substitution to find $f(4)$ and $f(-3)$ for each function.

5. $f(x) = 2x^3 + x^2 - 5x + 3$

6. $f(x) = 3x^3 - 4x + 2$

7. $f(x) = 5x^3 - 4x^2 + 2$

8. $f(x) = 2x^4 - 4x^3 + 3x^2 + x - 6$

9. $f(x) = 5x^4 + 3x^3 - 4x^2 - 2x + 4$

10. $f(x) = 3x^4 - 2x^3 - x^2 + 2x - 5$

11. $f(x) = 2x^4 - 4x^3 - x^2 - 6x + 3$

12. $f(x) = 4x^4 - 4x^3 + 3x^2 - 2x - 3$

6-6 Study Guide and Intervention *(continued)*

The Remainder and Factor Theorems

Factors of Polynomials The **Factor Theorem** can help you find all the factors of a polynomial.

Factor Theorem	The binomial $x - a$ is a factor of the polynomial $f(x)$ if and only if $f(a) = 0$.

Example Show that $x + 5$ is a factor of $x^3 + 2x^2 - 13x + 10$. Then find the remaining factors of the polynomial.

By the Factor Theorem, the binomial $x + 5$ is a factor of the polynomial if -5 is a zero of the polynomial function. To check this, use synthetic substitution.

$$
\begin{array}{r|rrrr}
-5 & 1 & 2 & -13 & 10 \\
   &   & -5 & 15 & -10 \\
\hline
   & 1 & -3 & 2 & 0
\end{array}
$$

Since the remainder is 0, $x + 5$ is a factor of the polynomial. The polynomial $x^3 + 2x^2 - 13x + 10$ can be factored as $(x + 5)(x^2 - 3x + 2)$. The depressed polynomial $x^2 - 3x + 2$ can be factored as $(x - 2)(x - 1)$.

So $x^3 + 2x^2 - 13x + 10 = (x + 5)(x - 2)(x - 1)$.

Exercises

Given a polynomial and one of its factors, find the remaining factors of the polynomial.

1. $x^3 + x^2 - 10x + 8;\ x - 2$

2. $x^3 - 4x^2 - 11x + 30;\ x + 3$

3. $x^3 + 15x^2 + 71x + 105;\ x + 7$

4. $x^3 - 7x^2 - 26x + 72;\ x + 4$

5. $2x^3 - x^2 - 7x + 6;\ x - 1$

6. $3x^3 - x^2 - 62x - 40;\ x + 4$

7. $12x^3 - 71x^2 + 57x - 10;\ x - 5$

8. $14x^3 + x^2 - 24x + 9;\ x - 1$

9. $x^3 + x + 10;\ x + 2$

10. $2x^3 - 11x^2 + 19x - 28;\ x - 4$

11. $3x^3 - 13x^2 - 34x + 24;\ x - 6$

12. $x^4 + x^3 - 11x^2 - 9x + 18;\ x - 1$

6-7 Study Guide and Intervention

Roots and Zeros

Synthetic Types of Roots The following statements are equivalent for any polynomial function $f(x)$.

- c is a zero of the polynomial function $f(x)$.
- c is a root or solution of the polynomial equation $f(x) = 0$.
- $(x - c)$ is a factor of the polynomial $f(x)$.
- If c is real, then $(c, 0)$ is an intercept of the graph of $f(x)$.

Fundamental Theorem of Algebra	Every polynomial equation with degree greater than zero has at least one root in the set of complex numbers.
Corollary to the Fundamental Theorem of Algebras	A polynomial equation of the form $P(x) = 0$ of degree n with complex coefficients has exactly n roots in the set of complex numbers, including repeated roots.
Descartes' Rule of Signs	If $P(x)$ is a polynomial with real coefficients whose terms are arranged in descending powers of the variable, • the number of positive real zeros of $y = P(x)$ is the same as the number of changes in sign of the coefficients of the terms, or is less than this by an even number, and • the number of negative real zeros of $y = P(x)$ is the same as the number of changes in sign of the coefficients of the terms of $P(-x)$, or is less than this number by an even number.

Example 1 Solve the equation $6x^3 + 3x = 0$. State the number and type of roots.

$6x^3 + 3x = 0$

$3x(2x^2 + 1) = 0$

Use the Zero Product Property.

$3x = 0$ or $2x^2 + 1 = 0$

$x = 0$ or $2x^2 = -1$

$x = \pm \dfrac{i\sqrt{2}}{2}$

The equation has one real root, 0, and two imaginary roots, $\pm \dfrac{i\sqrt{2}}{2}$.

Example 2 State the number of positive real zeros, negative real zeros, and imaginary zeros for $p(x) = 4x^4 - 3x^3 - x^2 + 2x - 5$.

Since $p(x)$ has degree 4, it has 4 zeros.

Since there are three sign changes, there are 3 or 1 positive real zeros.

Find $p(-x)$ and count the number of changes in sign for its coefficients.

$p(-x) = 4(-x)^4 - 3(-x)^3 + (-x)^2 + 2(-x) - 5$

$\quad\quad = 4x^4 + 3x^3 + x^2 - 2x - 5$

Since there is one sign change, there is exactly 1 negative real zero.

Thus, there are 3 positive and 1 negative real zero or 1 positive and 1 negative real zeros and 2 imaginary zeros.

Exercises

Solve each equation. State the number and type of roots.

1. $x^2 + 4x - 21 = 0$

2. $2x^3 - 50x = 0$

3. $12x^3 + 100x = 0$

State the possible number of positive real zeros, negative real zeros, and imaginary zeros for each function.

4. $f(x) = 3x^3 + x^2 - 8x - 12$

5. $f(x) = 3x^5 - x^4 - x^3 + 6x^2 - 5$

6-7 Study Guide and Intervention (continued)

Roots and Zeros

Find Zeros

Complex Conjugate Theorem	Suppose a and b are real numbers with $b \neq 0$. If $a + bi$ is a zero of a polynomial function with real coefficients, then $a - bi$ is also a zero of the function.

Example **Find all of the zeros of $f(x) = x^4 - 15x^2 + 38x - 60$.**

Since $f(x)$ has degree 4, the function has 4 zeros.

$f(x) = x^4 - 15x^2 + 38x - 60$ $f(-x) = x^4 - 15x^2 - 38x - 60$

Since there are 3 sign changes for the coefficients of $f(x)$, the function has 3 or 1 positive real zeros. Since there is + sign change for the coefficients of $f(-x)$, the function has 1 negative real zero. Use synthetic substitution to test some possible zeros.

$$
\begin{array}{r|rrrrr}
2 & 1 & 0 & -15 & 38 & -60 \\
  &   & 2 & 4 & -22 & 32 \\
\hline
  & 1 & 2 & -11 & 16 & -28 \\
\end{array}
$$

$$
\begin{array}{r|rrrrr}
3 & 1 & 0 & -15 & 38 & -60 \\
  &   & 3 & 9 & -18 & 60 \\
\hline
  & 1 & 3 & -6 & 20 & 0 \\
\end{array}
$$

So 3 is a zero of the polynomial function. Now try synthetic substitution again to find a zero of the depressed polynomial.

$$
\begin{array}{r|rrrr}
-2 & 1 & 3 & -6 & 20 \\
   &   & -2 & -2 & 16 \\
\hline
   & 1 & 1 & -8 & 36 \\
\end{array}
$$

$$
\begin{array}{r|rrrr}
-4 & 1 & 3 & -6 & 20 \\
   &   & -4 & 4 & 8 \\
\hline
   & 1 & -1 & -2 & 28 \\
\end{array}
$$

$$
\begin{array}{r|rrrr}
-5 & 1 & 3 & -6 & 20 \\
   &   & -5 & 10 & -20 \\
\hline
   & 1 & -2 & 4 & 0 \\
\end{array}
$$

So -5 is another zero. Use the Quadratic Formula on the depressed polynomial $x^2 - 2x + 4$ to find the other 1 zeros, $1 \pm i\sqrt{3}$.

The function has two real zeros at 3 and -5 and two imaginary zeros at $1 \pm i\sqrt{3}$.

Exercises

Find all zeros of each function.

1. $f(x) = x^3 + x^2 + 9x + 9$

2. $f(x) = x^3 - 3x^2 + 4x - 12$

3. $p(a) = a^3 - 10a^2 + 34a - 40$

4. $p(x) = x^3 - 5x^2 + 11x - 15$

5. $f(x) = x^3 + 6x + 20$

6. $f(x) = x^4 - 3x^3 + 21x^2 - 75x - 100$

6-8 Study Guide and Intervention

Rational Zero Theorem

Identify Rational Zeros

Rational Zero Theorem	Let $f(x) = a_nx^n + a_{n-1}x^{n-1} + \ldots + a_2x^2 + a_1x + a_0$ represent a polynomial function with integral coefficients. If $\frac{p}{q}$ is a rational number in simplest form and is a zero of $y = f(x)$, then p is a factor of a_0 and q is a factor of a_n.
Corollary (Integral Zero Theorem)	If the coefficients of a polynomial are integers such that $a_n = 1$ and $a_0 \neq 0$, any rational zeros of the function must be factors of a_0.

Example List all of the possible rational zeros of each function.

a. $f(x) = 3x^4 - 2x^2 + 6x - 10$

If $\frac{p}{q}$ is a rational root, then p is a factor of -10 and q is a factor of 3. The possible values for p are ± 1, ± 2, ± 5, and ± 10. The possible values for q are 61 and 63. So all of the possible rational zeros are $\frac{p}{q} = \pm 1$, ± 2, ± 5, ± 10, $\pm\frac{1}{3}$, $\pm\frac{2}{3}$, $\pm\frac{5}{3}$, and $\pm\frac{10}{3}$.

b. $q(x) = x^3 - 10x^2 + 14x - 36$

Since the coefficient of x^3 is 1, the possible rational zeros must be the factors of the constant term -36. So the possible rational zeros are ± 1, ± 2, ± 3, ± 4, ± 6, ± 9, ± 12, ± 18, and ± 36.

Exercises

List all of the possible rational zeros of each function.

1. $f(x) = x^3 + 3x^2 - x + 8$

2. $g(x) = x^5 - 7x^4 + 3x^2 + x - 20$

3. $h(x) = x^4 - 7x^3 - 4x^2 + x - 49$

4. $p(x) = 2x^4 - 5x^3 + 8x^2 + 3x - 5$

5. $q(x) = 3x^4 - 5x^3 + 10x + 12$

6. $r(x) = 4x^5 - 2x + 18$

7. $f(x) = x^7 - 6x^5 - 3x^4 + x^3 + 4x^2 - 120$

8. $g(x) = 5x^6 - 3x^4 + 5x^3 + 2x^2 - 15$

9. $h(x) = 6x^5 - 3x^4 + 12x^3 + 18x^2 - 9x + 21$

10. $p(x) = 2x^7 - 3x^6 + 11x^5 - 20x^2 + 11$

6-8 Study Guide and Intervention *(continued)*

Rational Zero Theorem

Find Rational Zeros

Example 1 **Find all of the rational zeros of $f(x) = 5x^3 + 12x^2 - 29x + 12$.**

From the corollary to the Fundamental Theorem of Algebra, we know that there are exactly 3 complex roots. According to Descartes' Rule of Signs there are 2 or 0 positive real roots and 1 negative real root. The possible rational zeros are $\pm 1, \pm 2, \pm 3, \pm 4, \pm 6, \pm 12,$ $\pm\frac{1}{5}, \pm\frac{2}{5}, \pm\frac{3}{5}, \pm\frac{4}{5}, \pm\frac{6}{5}, \pm\frac{12}{5}$. Make a table and test some possible rational zeros.

$\frac{p}{q}$	5	12	−29	12
1	5	17	−12	0

Since $f(1) = 0$, you know that $x = 1$ is a zero.

The depressed polynomial is $5x^2 + 17x - 12$, which can be factored as $(5x - 3)(x + 4)$.

By the Zero Product Property, this expression equals 0 when $x = \frac{3}{5}$ or $x = -4$.

The rational zeros of this function are $1, \frac{3}{5}$, and -4.

Example 2 **Find all of the zeros of $f(x) = 8x^4 + 2x^3 + 5x^2 + 2x - 3$.**

There are 4 complex roots, with 1 positive real root and 3 or 1 negative real roots. The possible rational zeros are $\pm 1, \pm 3, \pm\frac{1}{2}, \pm\frac{1}{4}, \pm\frac{1}{8}, \pm\frac{3}{2}, \pm\frac{3}{4}$, and $\pm\frac{3}{8}$.

Make a table and test some possible values.

$\frac{p}{q}$	8	2	5	2	−3
1	8	10	15	17	14
2	8	18	41	84	165
$\frac{1}{2}$	8	6	8	6	0

Since $f\left(\frac{1}{2}\right) = 0$, we know that $x = \frac{1}{2}$ is a root.

The depressed polynomial is $8x^3 + 6x^2 + 8x + 6$.

Try synthetic substitution again. Any remaining rational roots must be negative.

$\frac{p}{q}$	8	6	8	6
$-\frac{1}{4}$	8	4	7	$4\frac{1}{4}$
$-\frac{3}{4}$	8	0	8	0

$x = -\frac{3}{4}$ is another rational root.

The depressed polynomial is $8x^2 + 8 = 0$, which has roots $\pm i$.

The zeros of this function are $\frac{1}{2}, -\frac{3}{4}$, and $\pm i$.

Exercises

Find all of the rational zeros of each function.

1. $f(x) = x^3 + 4x^2 - 25x - 28$

2. $f(x) = x^3 + 6x^2 + 4x + 24$

Find all of the zeros of each function.

3. $f(x) = x^4 + 2x^3 - 11x^2 + 8x - 60$

4. $f(x) = 4x^4 + 5x^3 + 30x^2 + 45x - 54$

7-1 Study Guide and Intervention

Operations on Functions

Arithmetic Operations

Operations with Functions

Sum	$(f + g)(x) = f(x) + g(x)$
Difference	$(f - g)(x) = f(x) - g(x)$
Product	$(f \cdot g)(x) = f(x) \cdot g(x)$
Quotient	$\left(\dfrac{f}{g}\right)(x) = \dfrac{f(x)}{g(x)},\ g(x) \neq 0$

Example Find $(f + g)(x)$, $(f - g)(x)$, $(f \cdot g)(x)$, and $\left(\dfrac{f}{g}\right)(x)$ for $f(x) = x^2 + 3x - 4$ and $g(x) = 3x - 2$.

$$(f + g)(x) = f(x) + g(x) \qquad \text{Addition of functions}$$
$$= (x^2 + 3x - 4) + (3x - 2) \qquad f(x) = x^2 + 3x - 4,\ g(x) = 3x - 2$$
$$= x^2 + 6x - 6 \qquad \text{Simplify.}$$

$$(f - g)(x) = f(x) - g(x) \qquad \text{Subtraction of functions}$$
$$= (x^2 + 3x - 4) - (3x - 2) \qquad f(x) = x^2 + 3x - 4,\ g(x) = 3x - 2$$
$$= x^2 - 2 \qquad \text{Simplify.}$$

$$(f \cdot g)(x) = f(x) \cdot g(x) \qquad \text{Multiplication of functions}$$
$$= (x^2 + 3x - 4)(3x - 2) \qquad f(x) = x^2 + 3x - 4,\ g(x) = 3x - 2$$
$$= x^2(3x - 2) + 3x(3x - 2) - 4(3x - 2) \qquad \text{Distributive Property}$$
$$= 3x^3 - 2x^2 + 9x^2 - 6x - 12x + 8 \qquad \text{Distributive Property}$$
$$= 3x^3 + 7x^2 - 18x + 8 \qquad \text{Simplify.}$$

$$\left(\dfrac{f}{g}\right)(x) = \dfrac{f(x)}{g(x)} \qquad \text{Division of functions}$$
$$= \dfrac{x^2 + 3x - 4}{3x - 2},\ x \neq \dfrac{2}{3} \qquad f(x) = x^2 + 3x - 4 \text{ and } g(x) = 3x - 2$$

Exercises

Find $(f + g)(x)$, $(f - g)(x)$, $(f \cdot g)(x)$, and $\left(\dfrac{f}{g}\right)(x)$ for each $f(x)$ and $g(x)$.

1. $f(x) = 8x - 3;\ g(x) = 4x + 5$

2. $f(x) = x^2 + x - 6;\ g(x) = x - 2$

3. $f(x) = 3x^2 - x + 5;\ g(x) = 2x - 3$

4. $f(x) = 2x - 1;\ g(x) = 3x^2 + 11x - 4$

5. $f(x) = x^2 - 1;\ g(x) = \dfrac{1}{x + 1}$

7-1 Study Guide and Intervention *(continued)*

Operations on Functions

Composition of Functions Suppose f and g are functions such that the range of g is a subset of the domain of f. Then the composite function $f \circ g$ can be described by the equation $[f \circ g](x) = f[g(x)]$.

Example 1 For $f = \{(1, 2), (3, 3), (2, 4), (4, 1)\}$ and $g = \{(1, 3), (3, 4), (2, 2), (4, 1)\}$, find $f \circ g$ and $g \circ f$ if they exist.

$f[g(1)] = f(3) = 3$ $f[g(2)] = f(2) = 4$ $f[g(3)] = f(4) = 1$ $f[g(4)] = f(1) = 2,$

So $f \circ g = \{(1, 3), (2, 4), (3, 1), (4, 2)\}$

$g[f(1)] = g(2) = 2$ $g[f(2)] = g(4) = 1$ $g[f(3)] = g(3) = 4$ $g[f(4)] = g(1) = 3,$

So $g \circ f = \{(1, 2), (2, 1), (3, 4), (4, 3)\}$

Example 2 Find $[g \circ h](x)$ and $[h \circ g](x)$ for $g(x) = 3x - 4$ and $h(x) = x^2 - 1$.

$[g \circ h](x) = g[h(x)]$ $[h \circ g](x) = h[g(x)]$

$\qquad = g(x^2 - 1)$ $= h(3x - 4)$

$\qquad = 3(x^2 - 1) - 4$ $= (3x - 4)^2 - 1$

$\qquad = 3x^2 - 7$ $= 9x^2 - 24x + 16 - 1$

$\qquad\qquad\qquad\qquad\qquad\quad = 9x^2 - 24x + 15$

Exercises

For each pair of functions, find $f \circ g$ and $g \circ f$, if they exist.

1. $f = \{(-1, 2), (5, 6), (0, 9)\},$
$\quad g = \{(6, 0), (2, -1), (9, 5)\}$

2. $f = \{(5, -2), (9, 8), (-4, 3), (0, 4)\},$
$\quad g = \{(3, 7), (-2, 6), (4, -2), (8, 10)\}$

Find $[f \circ g](x)$ and $[g \circ f](x)$, if they exist.

3. $f(x) = 2x + 7; g(x) = -5x - 1$

4. $f(x) = x^2 - 1; g(x) = -4x^2$

5. $f(x) = x^2 + 2x; g(x) = x - 9$

6. $f(x) = 5x + 4; g(x) = 3 - x$

7-2 Study Guide and Intervention
Inverse Functions and Relations

Find Inverses

Inverse Relations	Two relations are inverse relations if and only if whenever one relation contains the element (a, b), the other relation contains the element (b, a).
Property of Inverse Functions	Suppose f and f^{-1} are inverse functions. Then $f(a) = b$ if and only if $f^{-1}(b) = a$.

Example Find the inverse of the function $f(x) = \frac{2}{5}x - \frac{1}{5}$. Then graph the function and its inverse.

Step 1 Replace $f(x)$ with y in the original equation.

$$f(x) = \frac{2}{5}x - \frac{1}{5} \quad \rightarrow \quad y = \frac{2}{5}x - \frac{1}{5}$$

Step 2 Interchange x and y.

$$x = \frac{2}{5}y - \frac{1}{5}$$

Step 3 Solve for y.

$$x = \frac{2}{5}y - \frac{1}{5} \qquad \text{Inverse of } y = \frac{2}{5}x - \frac{1}{5}$$
$$5x = 2y - 1 \qquad \text{Multiply each side by 5.}$$
$$5x + 1 = 2y \qquad \text{Add 1 to each side.}$$
$$\frac{1}{2}(5x + 1) = y \qquad \text{Divide each side by 2.}$$

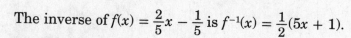

The inverse of $f(x) = \frac{2}{5}x - \frac{1}{5}$ is $f^{-1}(x) = \frac{1}{2}(5x + 1)$.

Exercises

Find the inverse of each function. Then graph the function and its inverse.

1. $f(x) = \frac{2}{3}x - 1$

2. $f(x) = 2x - 3$

3. $f(x) = \frac{1}{4}x - 2$

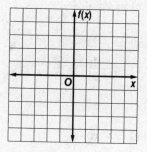

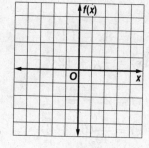

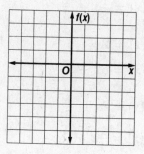

7-2 Study Guide and Intervention *(continued)*

Inverse Functions and Relations

Verifying Inverses

Inverse Functions	Two functions $f(x)$ and $g(x)$ are inverse functions if and only if $[f \circ g](x) = x$ and $[g \circ f](x) = x$.

Example 1 Determine whether $f(x) = 2x - 7$ and $g(x) = \frac{1}{2}(x + 7)$ are inverse functions.

$$[f \circ g](x) = f[g(x)]$$
$$= f\left[\frac{1}{2}(x + 7)\right]$$
$$= 2\left[\frac{1}{2}(x + 7)\right] - 7$$
$$= x + 7 - 7$$
$$= x$$

$$[g \circ f](x) = g[f(x)]$$
$$= g(2x - 7)$$
$$= \frac{1}{2}(2x - 7 + 7)$$
$$= x$$

The functions are inverses since both $[f \circ g](x) = x$ and $[g \circ f](x) = x$.

Example 2 Determine whether $f(x) = 4x + \frac{1}{3}$ and $g(x) = \frac{1}{4}x - 3$ are inverse functions.

$$[f \circ g](x) = f[g(x)]$$
$$= f\left(\frac{1}{4}x - 3\right)$$
$$= 4\left(\frac{1}{4}x - 3\right) + \frac{1}{3}$$
$$= x - 12 + \frac{1}{3}$$
$$= x - 11\frac{2}{3}$$

Since $[f \circ g](x) \neq x$, the functions are not inverses.

Exercises

Determine whether each pair of functions are inverse functions. Write *yes* or *no*.

1. $f(x) = 3x - 1$
$g(x) = \frac{1}{3}x + \frac{1}{3}$

2. $f(x) = \frac{1}{4}x + 5$
$g(x) = 4x - 20$

3. $f(x) = \frac{1}{2}x - 10$
$g(x) = 2x + \frac{1}{10}$

4. $f(x) = 2x + 5$
$g(x) = 5x + 2$

5. $f(x) = 8x - 12$
$g(x) = \frac{1}{8}x + 12$

6. $f(x) = -2x + 3$
$g(x) = -\frac{1}{2}x + \frac{3}{2}$

7. $f(x) = 4x - \frac{1}{2}$
$g(x) = \frac{1}{4}x + \frac{1}{8}$

8. $f(x) = 2x - \frac{3}{5}$
$g(x) = \frac{1}{10}(5x + 3)$

9. $f(x) = 4x + \frac{1}{2}$
$g(x) = \frac{1}{2}x - \frac{3}{2}$

10. $f(x) = 10 - \frac{x}{2}$
$g(x) = 20 - 2x$

11. $f(x) = 4x - \frac{4}{5}$
$g(x) = \frac{x}{4} + \frac{1}{5}$

12. $f(x) = 9 + \frac{3}{2}x$
$g(x) = \frac{2}{3}x - 6$

7-3 Study Guide and Intervention

Square Root Functions and Inequalities

Square Root Functions A function that contains the square root of a variable expression is a **square root function**. The domain of a square root function is those values for which the radicand is greater than or equal to 0.

Example Graph $y = \sqrt{3x - 2}$. State its domain and range.

Since the radicand cannot be negative, the domain of the function is $3x - 2 \geq 0$ or $x \geq \frac{2}{3}$.

The x-intercept is $\frac{2}{3}$. The range is $y \geq 0$.

Make a table of values and graph the function.

x	y
$\frac{2}{3}$	0
1	1
2	2
3	$\sqrt{7}$

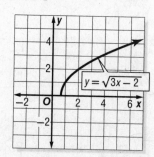

Exercises

Graph each function. State the domain and range.

1. $y = \sqrt{2x}$

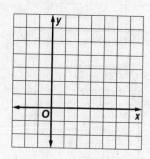

2. $y = -3\sqrt{x}$

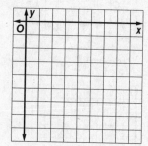

3. $y = -\sqrt{\dfrac{x}{2}}$

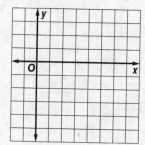

4. $y = 2\sqrt{x - 3}$

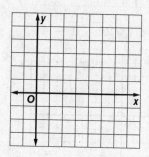

5. $y = -\sqrt{2x - 3}$

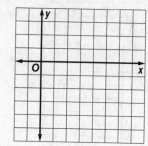

6. $y = \sqrt{2x + 5}$

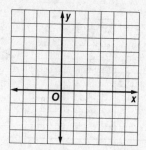

7-3 Study Guide and Intervention *(continued)*

Square Root Functions and Inequalities

Square Root Inequalities A **square root inequality** is an inequality that contains the square root of a variable expression. Use what you know about graphing square root functions and graphing inequalities to graph square root inequalities.

Example **Graph $y \leq \sqrt{2x - 1} + 2$.**

Graph the related equation $y = \sqrt{2x - 1} + 2$. Since the boundary should be included, the graph should be solid.

The domain includes values for $x \geq \frac{1}{2}$, so the graph is to the right of $x = \frac{1}{2}$.

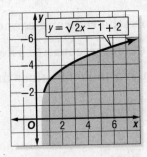

Exercises

Graph each inequality.

1. $y < 2\sqrt{x}$

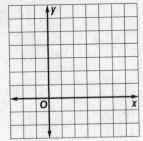

2. $y > \sqrt{x + 3}$

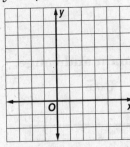

3. $y < 3\sqrt{2x - 1}$

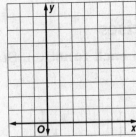

4. $y < \sqrt{3x - 4}$

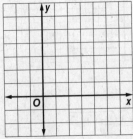

5. $y \geq \sqrt{x + 1} - 4$

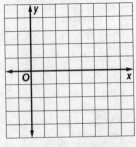

6. $y > 2\sqrt{2x - 3}$

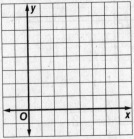

7. $y \geq \sqrt{3x + 1} - 2$

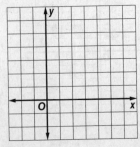

8. $y \leq \sqrt{4x - 2} + 1$

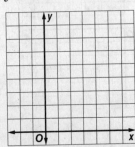

9. $y < 2\sqrt{2x - 1} - 4$

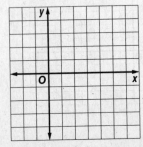

7-4 Study Guide and Intervention

nth Roots

Simplify Radicals

Square Root	For any real numbers a and b, if $a^2 = b$, then a is a square root of b.
nth Root	For any real numbers a and b, and any positive integer n, if $a^n = b$, then a is an nth root of b.
Real nth Roots of b, $\sqrt[n]{b}$, $-\sqrt[n]{b}$	1. If n is even and $b > 0$, then b has one positive real root and one real negative root.
	2. If n is odd and $b > 0$, then b has one positive real root.
	3. If n is even and $b < 0$, then b has no real roots.
	4. If n is odd and $b < 0$, then b has one negative real root.

Example 1 Simplify $\sqrt{49z^8}$.

$\sqrt{49z^8} = \sqrt{(7z^4)^2} = 7z^4$

z^4 must be positive, so there is no need to take the absolute value.

Example 2 Simplify $-\sqrt[3]{(2a-1)^6}$

$-\sqrt[3]{(2a-1)^6} = -\sqrt[3]{[(2a-1)^2]^3} = -(2a-1)^2$

Exercises

Simplify.

1. $\sqrt{81}$

2. $\sqrt[3]{-343}$

3. $\sqrt{144p^6}$

4. $\pm\sqrt{4a^{10}}$

5. $\sqrt[5]{243p^{10}}$

6. $-\sqrt[3]{m^6n^9}$

7. $\sqrt[3]{-b^{12}}$

8. $\sqrt{16a^{10}b^8}$

9. $\sqrt{121x^6}$

10. $\sqrt{(4k)^4}$

11. $\pm\sqrt{169r^4}$

12. $-\sqrt[3]{-27p^6}$

13. $-\sqrt{625y^2z^4}$

14. $\sqrt{36q^{34}}$

15. $\sqrt{100x^2y^4z^6}$

16. $\sqrt[3]{-0.027}$

17. $-\sqrt{-0.36}$

18. $\sqrt{0.64p^{10}}$

19. $\sqrt[4]{(2x)^8}$

20. $\sqrt{(11y^2)^4}$

21. $\sqrt[3]{(5a^2b)^6}$

22. $\sqrt{(3x-1)^2}$

23. $\sqrt[3]{(m-5)^6}$

24. $\sqrt{36x^2-12x+1}$

7-4 Study Guide and Intervention *(continued)*

*n*th Roots

Approximate Radicals with a Calculator

Irrational Number	a number that cannot be expressed as a terminating or a repeating decimal

Radicals such as $\sqrt{2}$ and $\sqrt{3}$ are examples of irrational numbers. Decimal approximations for irrational numbers are often used in applications. These approximations can be easily found with a calculator.

Example Use a calculator to approximate $\sqrt[5]{18.2}$ to three decimal places.

$\sqrt[5]{18.2} \approx 1.787$

Exercises

Use a calculator to approximate each value to three decimal places.

1. $\sqrt{62}$

2. $\sqrt{1050}$

3. $\sqrt[3]{0.054}$

4. $-\sqrt[4]{5.45}$

5. $\sqrt{5280}$

6. $\sqrt{18,600}$

7. $\sqrt{0.095}$

8. $\sqrt[3]{-15}$

9. $\sqrt[5]{100}$

10. $\sqrt[6]{856}$

11. $\sqrt{3200}$

12. $\sqrt{0.05}$

13. $\sqrt{12,500}$

14. $\sqrt{0.60}$

15. $-\sqrt[4]{500}$

16. $\sqrt[3]{0.15}$

17. $\sqrt[6]{4200}$

18. $\sqrt{75}$

19. **LAW ENFORCEMENT** The formula $r = 2\sqrt{5L}$ is used by police to estimate the speed r in miles per hour of a car if the length L of the car's skid mark is measures in feet. Estimate to the nearest tenth of a mile per hour the speed of a car that leaves a skid mark 300 feet long.

20. **SPACE TRAVEL** The distance to the horizon d miles from a satellite orbiting h miles above Earth can be approximated by $d = \sqrt{8000h + h^2}$. What is the distance to the horizon if a satellite is orbiting 150 miles above Earth?

7-5 Study Guide and Intervention

Operations with Radical Expressions

Simplify Radicals

Product Property of Radicals	For any real numbers a and b, and any integer $n > 1$: 1. if n is even and a and b are both nonnegative, then $\sqrt[n]{ab} = \sqrt[n]{a} \cdot \sqrt[n]{b}$. 2. if n is odd, then $\sqrt[n]{ab} = \sqrt[n]{a} \cdot \sqrt[n]{b}$.

To simplify a square root, follow these steps:

1. Factor the radicand into as many squares as possible.
2. Use the Product Property to isolate the perfect squares.
3. Simplify each radical.

Quotient Property of Radicals	For any real numbers a and $b \neq 0$, and any integer $n > 1$, $\sqrt[n]{\dfrac{a}{b}} = \dfrac{\sqrt[n]{a}}{\sqrt[n]{b}}$, if all roots are defined.

To eliminate radicals from a denominator or fractions from a radicand, multiply the numerator and denominator by a quantity so that the radicand has an exact root.

Example 1 Simplify $\sqrt[3]{-6a^5b^7}$.

$$\sqrt[3]{-16a^5b^7} = \sqrt[3]{(-2)^3 \cdot 2 \cdot a^3 \cdot a^2 \cdot (b^2)^3 \cdot b}$$
$$= -2ab^2 \sqrt[3]{2a^2b}$$

Example 2 Simplify $\sqrt{\dfrac{8x^3}{45y^5}}$.

$$\sqrt{\frac{8x^3}{45y^5}} = \sqrt{\frac{8x^3}{45y^5}} \qquad \text{Quotient Property}$$

$$= \frac{\sqrt{(2x)^2 \cdot 2x}}{\sqrt{(3y^2)^2 \cdot 5y}} \qquad \text{Factor into squares.}$$

$$= \frac{\sqrt{(2x)^2} \cdot \sqrt{2x}}{\sqrt{(3y^2)^2} \cdot \sqrt{5y}} \qquad \text{Product Property}$$

$$= \frac{2|x|\sqrt{2x}}{3y^2\sqrt{5y}} \qquad \text{Simplify.}$$

$$= \frac{2|x|\sqrt{2x}}{3y^2\sqrt{5y}} \cdot \frac{\sqrt{5y}}{\sqrt{5y}} \qquad \text{Rationalize the denominator.}$$

$$= \frac{2|x|\sqrt{10xy}}{15y^3} \qquad \text{Simplify.}$$

Exercises

Simplify.

1. $5\sqrt{54}$

2. $\sqrt[4]{32a^9b^{20}}$

3. $\sqrt{75x^4y^7}$

4. $\sqrt{\dfrac{36}{125}}$

5. $\sqrt{\dfrac{a^6b^3}{98}}$

6. $\sqrt[3]{\dfrac{p^5q^3}{40}}$

7-5 Study Guide and Intervention *(continued)*

Operations with Radical Expressions

Operations with Radicals When you add expressions containing radicals, you can add only like terms or **like radical expressions**. Two radical expressions are called *like radical expressions* if both the indices and the radicands are alike.

To multiply radicals, use the Product and Quotient Properties. For products of the form $(a\sqrt{b} + c\sqrt{d}) \cdot (e\sqrt{f} + g\sqrt{h})$, use the FOIL method. To rationalize denominators, use **conjugates**. Numbers of the form $a\sqrt{b} + c\sqrt{d}$ and $a\sqrt{b} - c\sqrt{d}$, where a, b, c, and d are rational numbers, are called conjugates. The product of conjugates is always a rational number.

Example 1 Simplify $2\sqrt{50} + 4\sqrt{500} - 6\sqrt{125}$.

$$\begin{aligned}
2\sqrt{50} + 4\sqrt{500} - 6\sqrt{125} &= 2\sqrt{5^2 \cdot 2} + 4\sqrt{10^2 \cdot 5} - 6\sqrt{5^2 \cdot 5} \quad &\text{Factor using squares.}\\
&= 2 \cdot 5 \cdot \sqrt{2} + 4 \cdot 10 \cdot \sqrt{5} - 6 \cdot 5 \cdot \sqrt{5} \quad &\text{Simplify square roots.}\\
&= 10\sqrt{2} + 40\sqrt{5} - 30\sqrt{5} \quad &\text{Multiply.}\\
&= 10\sqrt{2} + 10\sqrt{5} \quad &\text{Combine like radicals.}
\end{aligned}$$

Example 2 Simplify $(2\sqrt{3} - 4\sqrt{2})(\sqrt{3} + 2\sqrt{2})$.

$$\begin{aligned}
&(2\sqrt{3} - 4\sqrt{2})(\sqrt{3} + 2\sqrt{2})\\
&= 2\sqrt{3} \cdot \sqrt{3} + 2\sqrt{3} \cdot 2\sqrt{2} - 4\sqrt{2} \cdot \sqrt{3} - 4\sqrt{2} \cdot 2\sqrt{2}\\
&= 6 + 4\sqrt{6} - 4\sqrt{6} - 16\\
&= -10
\end{aligned}$$

Example 3 Simplify $\dfrac{2 - \sqrt{5}}{3 + \sqrt{5}}$.

$$\begin{aligned}
\frac{2 - \sqrt{5}}{3 + \sqrt{5}} &= \frac{2 - \sqrt{5}}{3 + \sqrt{5}} \cdot \frac{3 - \sqrt{5}}{3 - \sqrt{5}}\\
&= \frac{6 - 2\sqrt{5} - 3\sqrt{5} + (\sqrt{5})^2}{3^2 - (\sqrt{5})^2}\\
&= \frac{6 - 5\sqrt{5} + 5}{9 - 5}\\
&= \frac{11 - 5\sqrt{5}}{4}
\end{aligned}$$

Exercises

Simplify.

1. $3\sqrt{2} + \sqrt{50} - 4\sqrt{8}$

2. $\sqrt{20} + \sqrt{125} - \sqrt{45}$

3. $\sqrt{300} - \sqrt{27} - \sqrt{75}$

4. $\sqrt[3]{81} \cdot \sqrt[3]{24}$

5. $\sqrt[3]{2}\left(\sqrt[3]{4} + \sqrt[3]{12}\right)$

6. $2\sqrt{3}(\sqrt{15} + \sqrt{60})$

7. $(2 + 3\sqrt{7})(4 + \sqrt{7})$

8. $(6\sqrt{3} - 4\sqrt{2})(3\sqrt{3} + \sqrt{2})$

9. $(4\sqrt{2} - 3\sqrt{5})(2\sqrt{20} + 5)$

10. $\dfrac{5\sqrt{48} + \sqrt{75}}{5\sqrt{3}}$

11. $\dfrac{4 + \sqrt{2}}{2 - \sqrt{2}}$

12. $\dfrac{5 - 3\sqrt{3}}{1 + 2\sqrt{3}}$

7-6 Study Guide and Intervention

Rational Exponents

Rational Exponents and Radicals

Definition of $b^{\frac{1}{n}}$	For any real number b and any positive integer n, $b^{\frac{1}{n}} = \sqrt[n]{b}$, except when $b < 0$ and n is even.
Definition of $b^{\frac{m}{n}}$	For any nonzero real number b, and any integers m and n, with $n > 1$, $b^{\frac{m}{n}} = \sqrt[n]{b^m} = (\sqrt[n]{b})^m$, except when $b < 0$ and n is even.

Example 1 Write $28^{\frac{1}{2}}$ in radical form.

Notice that $28 > 0$.

$$28^{\frac{1}{2}} = \sqrt{28}$$
$$= \sqrt{2^2 \cdot 7}$$
$$= \sqrt{2^2} \cdot \sqrt{7}$$
$$= 2\sqrt{7}$$

Example 2 Evaluate $\left(\frac{-8}{-125}\right)^{\frac{1}{3}}$.

Notice that $-8 < 0$, $-125 < 0$, and 3 is odd.

$$\left(\frac{-8}{-125}\right)^{\frac{1}{3}} = \frac{\sqrt[3]{-8}}{\sqrt[3]{-125}}$$
$$= \frac{-2}{-5}$$
$$= \frac{2}{5}$$

Exercises

Write each expression in radical form, or write each radical in exponential form.

1. $11^{\frac{1}{7}}$

2. $15^{\frac{1}{3}}$

3. $300^{\frac{3}{2}}$

4. $\sqrt{47}$

5. $\sqrt[3]{3a^5b^2}$

6. $\sqrt[4]{162p^5}$

Evaluate each expression.

7. $-27^{\frac{2}{3}}$

8. $216^{\frac{1}{3}}$

9. $(0.0004)^{\frac{1}{2}}$

7-6 Study Guide and Intervention *(continued)*

Rational Exponents

Simplify Expressions All the properties of powers from Lesson 6-1 apply to rational exponents. When you simplify expressions with rational exponents, leave the exponent in rational form, and write the expression with all positive exponents. Any exponents in the denominator must be positive integers.

When you simplify radical expressions, you may use rational exponents to simplify, but your answer should be in radical form. Use the smallest index possible.

Example 1 **Simplify $y^{\frac{2}{3}} \cdot y^{\frac{3}{8}}$.**

$y^{\frac{2}{3}} \cdot y^{\frac{3}{8}} = y^{\frac{2}{3}+\frac{3}{8}} = y^{\frac{25}{24}}$

Example 2 **Simplify $\sqrt[4]{144x^6}$.**

$\sqrt[4]{144x^6} = (144x^6)^{\frac{1}{4}}$

$= (2^4 \cdot 3^2 \cdot x^6)^{\frac{1}{4}}$

$= (2^4)^{\frac{1}{4}} \cdot (3^2)^{\frac{1}{4}} \cdot (x^6)^{\frac{1}{4}}$

$= 2 \cdot 3^{\frac{1}{2}} \cdot x^{\frac{3}{2}} = 2x \cdot (3x)^{\frac{1}{2}} = 2x\sqrt{3x}$

Exercises

Simplify each expression.

1. $x^{\frac{4}{5}} \cdot x^{\frac{6}{5}}$

2. $\left(y^{\frac{2}{3}}\right)^{\frac{3}{4}}$

3. $p^{\frac{4}{5}} \cdot p^{\frac{7}{10}}$

4. $\left(m^{\frac{6}{5}}\right)^{\frac{2}{5}}$

5. $x^{\frac{3}{8}} \cdot x^{\frac{4}{3}}$

6. $\left(s^{\frac{1}{6}}\right)^{\frac{4}{3}}$

7. $\dfrac{p}{p^{\frac{1}{3}}}$

8. $\dfrac{x^{\frac{1}{2}}}{x^{\frac{1}{3}}}$

9. $\sqrt[6]{128}$

10. $\sqrt[4]{49}$

11. $\sqrt[5]{288}$

12. $\sqrt{32} \cdot 3\sqrt{16}$

13. $\sqrt[3]{25} \cdot \sqrt{125}$

14. $\sqrt[6]{16}$

15. $\dfrac{a\sqrt[3]{b^4}}{\sqrt{ab^3}}$

7-7 Study Guide and Intervention

Solving Radical Equations and Inequalities

Solve Radical Equations The following steps are used in solving equations that have variables in the radicand. Some algebraic procedures may be needed before you use these steps.

Step 1	Isolate the radical on one side of the equation.
Step 2	To eliminate the radical, raise each side of the equation to a power equal to the index of the radical.
Step 3	Solve the resulting equation.
Step 4	Check your solution in the original equation to make sure that you have not obtained any extraneous roots.

Example 1 Solve $2\sqrt{4x + 8} - 4 = 8$.

$2\sqrt{4x + 8} - 4 = 8$	Original equation
$2\sqrt{4x + 8} = 12$	Add 4 to each side.
$\sqrt{4x + 8} = 6$	Isolate the radical.
$4x + 8 = 36$	Square each side.
$4x = 28$	Subtract 8 from each side.
$x = 7$	Divide each side by 4.

Check

$2\sqrt{4(7) + 8} - 4 \overset{?}{=} 8$

$2\sqrt{36} - 4 \overset{?}{=} 8$

$2(6) - 4 \overset{?}{=} 8$

$8 = 8$

The solution $x = 7$ checks.

Example 2 Solve $\sqrt{3x + 1} = \sqrt{5x} - 1$.

$\sqrt{3x + 1} = \sqrt{5x} - 1$	Original equation
$3x + 1 = 5x - 2\sqrt{5x} + 1$	Square each side.
$2\sqrt{5x} = 2x$	Simplify.
$\sqrt{5x} = x$	Isolate the radical.
$5x = x^2$	Square each side.
$x^2 - 5x = 0$	Subtract 5x from each side.
$x(x - 5) = 0$	Factor.
$x = 0$ or $x = 5$	

Check

$\sqrt{3(0) + 1} = 1$, but $\sqrt{5(0)} - 1 = -1$, so 0 is not a solution.

$\sqrt{3(5) + 1} = 4$, and $\sqrt{5(5)} - 1 = 4$, so the solution is $x = 5$.

Exercises

Solve each equation.

1. $3 + 2x\sqrt{3} = 5$

2. $2\sqrt{3x + 4} + 1 = 15$

3. $8 + \sqrt{x + 1} = 2$

4. $\sqrt{5 - x} - 4 = 6$

5. $12 + \sqrt{2x - 1} = 4$

6. $\sqrt{12 - x} = 0$

7. $\sqrt{21} - \sqrt{5x - 4} = 0$

8. $10 - \sqrt{2x} = 5$

9. $\sqrt{4 + 7x} = \sqrt{7x - 9}$

10. $4\sqrt[3]{2x + 11} - 2 = 10$

11. $2\sqrt{x - 11} = \sqrt{x + 4}$

12. $(9x - 11)^{\frac{1}{2}} = x + 1$

7-7　Study Guide and Intervention　(continued)

Solving Radical Equations and Inequalities

Solve Radical Inequalities A **radical inequality** is an inequality that has a variable in a radicand. Use the following steps to solve radical inequalities.

> **Step 1** If the index of the root is even, identify the values of the variable for which the radicand is nonnegative.
> **Step 2** Solve the inequality algebraically.
> **Step 3** Test values to check your solution.

Example　Solve $5 - \sqrt{20x + 4} \geq -3$.

Since the radicand of a square root must be greater than or equal to zero, first solve $20x + 4 \geq 0$.

$20x + 4 \geq 0$

$20x \geq -4$

$x \geq -\dfrac{1}{5}$

Now solve $5 - \sqrt{20x + 4} \geq -3$.

$5 - \sqrt{20x + 4} \geq -3$	Original inequality
$\sqrt{20x + 4} \leq 8$	Isolate the radical.
$20x + 4 \leq 64$	Eliminate the radical by squaring each side.
$20x \leq 60$	Subtract 4 from each side.
$x \leq 3$	Divide each side by 20.

It appears that $-\dfrac{1}{5} \leq x \leq 3$ is the solution. Test some values.

$x = -1$	$x = 0$	$x = 4$
$\sqrt{20(-1) + 4}$ is not a real number, so the inequality is not satisfied.	$5 - \sqrt{20(0) + 4} = 3$, so the inequality is satisfied.	$5 - \sqrt{20(4) + 4} \approx -4.2$, so the inequality is not satisfied.

Therefore the solution $-\dfrac{1}{5} \leq x \leq 3$ checks.

Exercises

Solve each inequality.

1. $\sqrt{c - 2} + 4 \geq 7$

2. $3\sqrt{2x - 1} + 6 < 15$

3. $\sqrt{10x + 9} - 2 > 5$

4. $8 - \sqrt{3x + 4} \geq 3$

5. $\sqrt{2x + 8} - 4 > 2$

6. $9 - \sqrt{6x + 3} \geq 6$

7. $2\sqrt{5x - 6} - 1 < 5$

8. $\sqrt{2x + 12} + 4 \geq 12$

9. $\sqrt{2d + 1} + \sqrt{d} \leq 5$

8-1 Study Guide and Intervention

Graphing Exponential Functions

Exponential Growth An **exponential growth function** has the form $y = b^x$, where $b > 1$. The graphs of exponential equations can be transformed by changing the value of the constants a, h, and k in the exponential equation: $f(x) = ab^{x-h} + k$.

Parent Function of Exponential Growth Functions, $f(x) = b^x, b > 1$	1. The function is continuous, one-to-one, and increasing. 2. The domain is the set of all real numbers. 3. The x-axis is the asymptote of the graph. 4. The range is the set of all non-zero real numbers. 5. The graph contains the point (0, 1).

Example   **Graph** $y = 4^x + 2$. **State the domain and range.**

Make a table of values. Connect the points to form a smooth curve.

x	−1	0	1	2	3
y	2.25	3	6	18	66

The domain of the function is all real numbers, while the range is the set of all positive real numbers greater than 2.

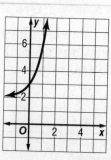

Exercises

Graph each function. State the domain and range.

1. $y = 3(2)^x$

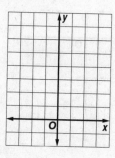

2. $y = \frac{1}{3}(3)^x$

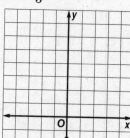

3. $y = 0.25(5)^x$

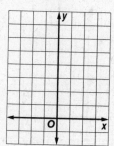

4. $y = 2(3)^x$

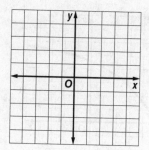

5. $y = 4^x - 2$

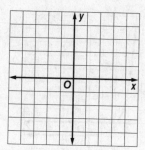

6. $y = 2^{x+5}$

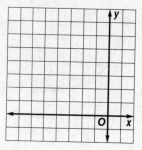

8-1 Study Guide and Intervention (continued)

Graphing Exponential Functions

Exponential Decay The following table summarizes the characteristics of **exponential decay** functions.

Parent Function of Exponential Decay Functions, $f(x) = b^x, 0 < b < 1$	1. The function is continuous, one-to-one, and decreasing. 2. The domain is the set of all real numbers. 3. The x-axis is the asymptote of the graph. 4. The range is the set of all positive real numbers. 5. The graph contains the point (0, 1).

Example Graph $y = \left(\frac{1}{2}\right)^x$. **State the domain and range.**

Make a table of values. Connect the points to form a smooth curve. The domain is all real numbers and the range is the set of all positive real numbers.

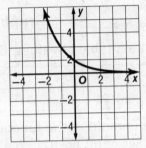

x	−2	−1	0	1	2
y	4	2	1	0.5	0.25

Exercises

Graph each function. State the domain and range.

1. $y = 6\left(\frac{1}{2}\right)^x$

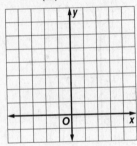

2. $y = -2\left(\frac{1}{4}\right)^x$

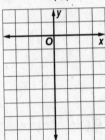

3. $y = -0.4(0.2)^x$

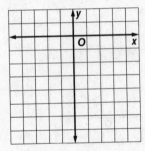

4. $y = \left(\frac{2}{5}\right)\left(\frac{1}{2}\right)^{x-1} + 2$

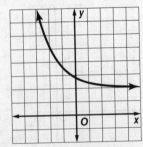

5. $y = 4\left(\frac{1}{5}\right)^{x+3} - 1$

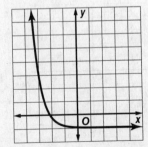

6. $y = \left(-\frac{1}{3}\right)\left(\frac{3}{4}\right)^{x-5} + 6$

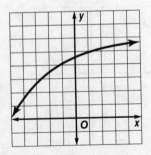

8-2 Study Guide and Intervention

Solving Exponential Equations and Inequalities

Solve Exponential Equations All the properties of rational exponents that you know also apply to real exponents. Remember that $a^m \cdot a^n = a^{m+n}$, $(a^m)^n = a^{mn}$, and $a^m \div a^n = a^{m-n}$.

Property of Equality for Exponential Functions	If b is a positive number other than 1, then $b^x = b^y$ if and only if $x = y$.

Example 1 Solve $4^{x-1} = 2^{x+5}$.

$4^{x-1} = 2^{x+5}$	Original equation
$(2^2)^{x-1} = 2^{x+5}$	Rewrite 4 as 2^2.
$2(x-1) = x+5$	Prop. of Inequality for Exponential Functions
$2x - 2 = x + 5$	Distributive Property
$x = 7$	Subtract x and add 2 to each side.

Example 2 Write an exponential function whose graph passes through the points (0, 3) and (4, 81).

The y-intercept is (0, 3), so $a = 3$. Since the other point is (4, 81), $b = \sqrt[4]{\dfrac{81}{3}}$.

Simplifying $\sqrt[4]{\dfrac{81}{3}} = \sqrt[4]{27} \approx 2.280$, the equation is $y = 3(2.280)^x$.

Exercises

Solve each equation.

1. $3^{2x-1} = 3^{x+2}$

2. $2^{3x} = 4^{x+2}$

3. $3^{2x-1} = \dfrac{1}{9}$

4. $4^{x+1} = 8^{2x+3}$

5. $8^{x-2} = \dfrac{1}{16}$

6. $25^{2x} = 125^{x+2}$

7. $9^{x+1} = 27^{x+4}$

8. $36^{2x+4} = 216^{x+5}$

9. $\left(\dfrac{1}{64}\right)^{x-2} = 16^{3x+1}$

Write an exponential function for the graph that passes through the given points.

10. (0, 4) and (2, 36)

11. (0, 6) and (1, 81)

12. (0, 5) and (6, 320)

13. (0, 2) and (5, 486)

14. (0, 8) and $\left(3, \dfrac{27}{8}\right)$

15. (0, 1) and (4, 625)

16. (0, 3) and (3, 24)

17. (0, 12) and (4, 144)

18. (0, 9) and (2, 49)

8-2 Study Guide and Intervention (continued)

Solve Exponential Equations and Inequalities

Solve Exponential Inequalities An **exponential inequality** is an inequality involving exponential functions.

Property of Inequality for Exponential Functions	If $b > 1$ then $b^x > b^y$ if and only if $x > y$ and $b^x < b^y$ if and only if $x < y$.

Example Solve $5^{2x-1} > \dfrac{1}{125}$.

$5^{2x-1} > \dfrac{1}{125}$ Original inequality

$5^{2x-1} > 5^{-3}$ Rewrite $\dfrac{1}{125}$ as 5^{-3}.

$2x - 1 > -3$ Prop. of Inequality for Exponential Functions

$2x > -2$ Add 1 to each side.

$x > -1$ Divide each sidet by 2.

The solution set is $\{x \mid x > -1\}$.

Exercises

Solve each inequality.

1. $3^{x-4} < \dfrac{1}{27}$

2. $4^{2x-2} > 2^{x+1}$

3. $5^{2x} < 125^{x-5}$

4. $10^{4x+1} > 100^{x-2}$

5. $7^{3x} < 49^{1-x}$

6. $8^{2x-5} < 4^{x+8}$

7. $16 \geq 4^{x+5}$

8. $\left(\dfrac{1}{27}\right)^{2x+1} \leq \left(\dfrac{1}{243}\right)^{3x-2}$

9. $\left(\dfrac{1}{2}\right)^{x-3} > 8^{2x}$

10. $\dfrac{1}{81} < 9^{2x-4}$

11. $32^{3x-4} > 128^{4x+3}$

12. $27^{2x-5} < \left(\dfrac{1}{9}\right)^{5x}$

13. $\left(\dfrac{1}{25}\right)^{2x-1} \leq 125^{3x+1}$

14. $\left(\dfrac{7}{343}\right)^{x-3} \geq \left(\dfrac{1}{49}\right)^{2x+1}$

15. $\left(\dfrac{9}{27}\right)^{6x-1} \geq \left(\dfrac{27}{9}\right)^{-x+6}$

8-3 Study Guide and Intervention

Logarithms and Logarithmic Functions

Logarithmic Functions and Expressions

Definition of Logarithm with Base b	Let b and x be positive numbers, $b \neq 1$. The logarithm of x with base b is denoted $\log_b x$ and is defined as the exponent y that makes the equation $b^y = x$ true.

The inverse of the exponential function $y = b^x$ is the **logarithmic function** $x = b^y$. This function is usually written as $y = \log_b x$.

Example 1 Write an exponential equation equivalent to $\log_3 243 = 5$.

$3^5 = 243$

Example 2 Write a logarithmic equation equivalent to $6^{-3} = \dfrac{1}{216}$.

$\log_6 \dfrac{1}{216} = -3$

Example 3 Evaluate $\log_8 16$.

$8^{\frac{4}{3}} = 16$, so $\log_8 16 = \dfrac{4}{3}$.

Exercises

Write each equation in exponential form.

1. $\log_{15} 225 = 2$

2. $\log_3 \dfrac{1}{27} = -3$

3. $\log_4 32 = \dfrac{5}{2}$

Write each equation in logarithmic form.

4. $2^7 = 128$

5. $3^{-4} = \dfrac{1}{81}$

6. $\left(\dfrac{1}{7}\right)^3 = \dfrac{1}{343}$

7. $7^{-2} = \dfrac{1}{49}$

8. $2^9 = 512$

9. $64^{\frac{2}{3}} = 16$

Evaluate each expression.

10. $\log_4 64$

11. $\log_2 64$

12. $\log_{100} 100{,}000$

13. $\log_5 625$

14. $\log_{27} 81$

15. $\log_{25} 5$

16. $\log_2 \dfrac{1}{128}$

17. $\log_{10} 0.00001$

18. $\log_4 \dfrac{1}{32}$

8-3 Study Guide and Intervention *(continued)*

Logarithms of Logarithmic Functions

Graphing Logarithmic Functions The function $y = \log_b x$, where $b \neq 1$, is called a **logarithmic function**. The graph of $f(x) = \log_b x$ represents a parent graph of the logarithmic functions. Properties of the parent function are described in the following table.

Parent function of Logarithmic Functions, $f(x) = \log_b x$	1. The function is continuous and one-to-one.
	2. The domain is the set of all positive real numbers.
	3. The y-axis is an asymptote of the graph.
	4. The range is the set of all real numbers.
	5. The graph contains the point (1, 0).

The graphs of logarithmic functions can be transformed by changing the value of the constants a, h, and k in the equation $f(x) = a \log_b (x - h) + k$.

Example **Graph $f(x) = -3 \log_{10} (x - 2) + 1$.**

This is a transformation of the graph of $f(x) = \log_{10} x$.

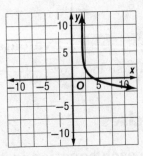

- $|a| = 3$: The graph expands vertically.

- $a < 0$: The graph is reflected across the x-axis.

- $h = 2$: The graph is translated 2 units to the right.

- $k = 1$: The graph is translated 1 unit up.

Exercises

Graph each function.

1. $f(x) = 4 \log_2 x$

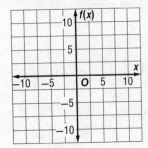

2. $f(x) = 4 \log_3 (x - 1)$

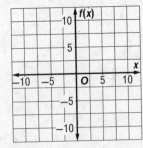

3. $f(x) = 2 \log_4 (x + 3) - 2$

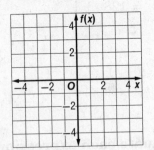

8-4 Study Guide and Intervention

Solving Logarithmic Equations and Inequalities

Solving Logarithmic Equations

Property of Equality for Logarithmic Functions	If b is a positive number other than 1, then $\log_b x = \log_b y$ if and only if $x = y$.

Example 1 Solve $\log_2 2x = 3$.

$\log_2 2x = 3$ Original equation

$2x = 2^3$ Definition of logarithm

$2x = 8$ Simplify.

$x = 4$ Simplify.

The solution is $x = 4$.

Example 2 Solve the equation $\log_2 (x + 17) = \log_2 (3x + 23)$.

Since the bases of the logarithms are equal, $(x + 17)$ must equal $(3x + 23)$.

$(x + 17) = (3x + 23)$

$-6 = 2x$

$x = -3$

Exercises

Solve each equation.

1. $\log_2 32 = 3x$

2. $\log_3 2c = -2$

3. $\log_{2x} 16 = -2$

4. $\log_{25} \left(\dfrac{x}{2}\right) = \dfrac{1}{2}$

5. $\log_4 (5x + 1) = 2$

6. $\log_8 (x - 5) = \dfrac{2}{3}$

7. $\log_4 (3x - 1) = \log_4 (2x + 3)$

8. $\log_2 (x^2 - 6) = \log_2 (2x + 2)$

9. $\log_{x + 4} 27 = 3$

10. $\log_2 (x + 3) = 4$

11. $\log_x 1000 = 3$

12. $\log_8 (4x + 4) = 2$

13. $\log_2 x = \log_2 12$

14. $\log_3 (x - 5) = \log_3 13$

15. $\log_{10} x = \log_{10} (5x - 20)$

16. $\log_5 x = \log_5 (2x - 1)$

17. $\log_4 (x + 12) = \log_4 4x$

18. $\log_6 (x - 3) = \log_6 2x$

8-4 Study Guide and Intervention *(continued)*

Solving Logarithmic Equations and Inequalities

Solving Logarithmic Inequalities

Property of Inequality for Logarithmic Functions	If $b > 1$, $x > 0$, and $\log_b x > y$, then $x > b^y$.
	If $b > 1$, $x > 0$, and $\log_b x < y$, then $0 < x < b^y$.
	If $b > 1$, then $\log_b x > \log_b y$ if and only if $x > y$,
	and $\log_b x < \log_b y$ if and only if $x < y$.

Example 1 Solve $\log_5 (4x - 3) < 3$.

$\log_5 (4x - 3) < 3$ Original equation

$0 < 4x - 3 < 5^3$ Property of Inequality

$3 < 4x < 125 + 3$ Simplify.

$\dfrac{3}{4} < x < 32$ Simplify.

The solution set is $\left\{ x \,\middle|\, \dfrac{3}{4} < x < 32 \right\}$.

Example 2 Solve the inequality $\log_3(3x - 4) < \log_3 (x + 1)$.

Since the base of the logarithms are equal to or greater than 1, $3x - 4 < x + 1$.

$2x < 5$

$x < \dfrac{5}{2}$

Since $3x - 4$ and $x + 1$ must both be positive numbers, solve $3x - 4 = 0$ for the lower bound of the inequality.

The solution is $\left\{ x \,\middle|\, \dfrac{4}{3} < x < \dfrac{5}{2} \right\}$.

Exercises

Solve each inequality.

1. $\log_2 2x > 2$

2. $\log_5 x > 2$

3. $\log_2 (3x + 1) < 4$

4. $\log_4 2x > -\dfrac{1}{2}$

5. $\log_3 (x + 3) < 3$

6. $\log_{27} 6x > \dfrac{2}{3}$

7. $\log_{10} 5x < \log_{10} 30$

8. $\log_{10} x < \log_{10} (2x - 4)$

9. $\log_{10} 3x < \log_{10} (7x - 8)$

10. $\log_2 (8x + 5) > \log_2 (9x - 18)$

11. $\log_{10} (3x + 7) < \log_{10} (7x - 3)$

12. $\log_2 (3x - 4) < \log_2 (2x + 7)$

8-5 Study Guide and Intervention

Properties of Logarithms

Properties of Logarithms Properties of exponents can be used to develop the following properties of logarithms.

Product Property of Logarithms	For all positive numbers a, b, and x, where $x \neq 1$, $\log_x ab = \log_x a + \log_x b$.
Quotient Property of Logarithms	For all positive numbers a, b, and x, where $x \neq 1$, $\log_x \frac{a}{b} = \log_x a - \log_x b$.
Power Property of Logarithms	For any real number p and positive numbers m and b, where $b \neq 1$, $\log_b m^p = p \log_b m$.

Example Use $\log_3 28 \approx 3.0331$ and $\log_3 4 \approx 1.2619$ to approximate the value of each expression.

a. $\log_3 36$

$\log_3 36 = \log_3 (3^2 \cdot 4)$
$\quad = \log_3 3^2 + \log_3 4$
$\quad = 2 + \log_3 4$
$\quad \approx 2 + 1.2619$
$\quad \approx 3.2619$

b. $\log_3 7$

$\log_3 7 = \log_3 \left(\frac{28}{4} \right)$
$\quad = \log_3 28 - \log_3 4$
$\quad \approx 3.0331 - 1.2619$
$\quad \approx 1.7712$

c. $\log_3 256$

$\log_3 256 = \log_3 (4^4)$
$\quad = 4 \cdot \log_3 4$
$\quad \approx 4(1.2619)$
$\quad \approx 5.0476$

Exercises

Use $\log_{12} 3 \approx 0.4421$ and $\log_{12} 7 \approx 0.7831$ to approximate the value of each expression.

1. $\log_{12} 21$

2. $\log_{12} \frac{7}{3}$

3. $\log_{12} 49$

4. $\log_{12} 36$

5. $\log_{12} 63$

6. $\log_{12} \frac{27}{49}$

7. $\log_{12} \frac{81}{49}$

8. $\log_{12} 16{,}807$

9. $\log_{12} 441$

Use $\log_5 3 \approx 0.6826$ and $\log_5 4 \approx 0.8614$ to approximate the value of each expression.

10. $\log_5 12$

11. $\log_5 100$

12. $\log_5 0.75$

13. $\log_5 144$

14. $\log_5 \frac{27}{16}$

15. $\log_5 375$

16. $\log_5 1.\overline{3}$

17. $\log_5 \frac{9}{16}$

18. $\log_5 \frac{81}{5}$

8-5 Study Guide and Intervention (continued)

Properties of Logarithms

Solve Logarithmic Equations You can use the properties of logarithms to solve equations involving logarithms.

> **Example** Solve each equation.

a. $2 \log_3 x - \log_3 4 = \log_3 25$

$2 \log_3 x - \log_3 4 = \log_3 25$	Original equation
$\log_3 x^2 - \log_3 4 = \log_3 25$	Power Property
$\log_3 \dfrac{x^2}{4} = \log_3 25$	Quotient Property
$\dfrac{x^2}{4} = 25$	Property of Equality for Logarithmic Functions
$x^2 = 100$	Multiply each side by 4.
$x = \pm 10$	Take the square root of each side.

Since logarithms are undefined for $x < 0$, -10 is an extraneous solution. The only solution is 10.

b. $\log_2 x + \log_2 (x + 2) = 3$

$\log_2 x + \log_2 (x + 2) = 3$	Original equation
$\log_2 x(x + 2) = 3$	Product Property
$x(x + 2) = 2^3$	Definition of logarithm
$x^2 + 2x = 8$	Distributive Property
$x^2 + 2x - 8 = 0$	Subtract 8 from each side.
$(x + 4)(x - 2) = 0$	Factor.
$x = 2$ or $x = -4$	Zero Product Property

Since logarithms are undefined for $x < 0$, -4 is an extraneous solution. The only solution is 2.

Exercises

Solve each equation. Check your solutions.

1. $\log_5 4 + \log_5 2x = \log_5 24$

2. $3 \log_4 6 - \log_4 8 = \log_4 x$

3. $\frac{1}{2} \log_6 25 + \log_6 x = \log_6 20$

4. $\log_2 4 - \log_2 (x + 3) = \log_2 8$

5. $\log_6 2x - \log_6 3 = \log_6 (x - 1)$

6. $2 \log_4 (x + 1) = \log_4 (11 - x)$

7. $\log_2 x - 3 \log_2 5 = 2 \log_2 10$

8. $3 \log_2 x - 2 \log_2 5x = 2$

9. $\log_3 (c + 3) - \log_3 (4c - 1) = \log_3 5$

10. $\log_5 (x + 3) - \log_5 (2x - 1) = 2$

8-6 Study Guide and Intervention

Common Logarithms

Common Logarithms Base 10 logarithms are called **common logarithms**. The expression $\log_{10} x$ is usually written without the subscript as $\log x$. Use the [LOG] key on your calculator to evaluate common logarithms.

The relation between exponents and logarithms gives the following identity.

Inverse Property of Logarithms and Exponents	$10^{\log x} = x$

Example 1   **Evaluate log 50 to the nearest ten-thousandth.**

Use the [LOG] key on your calculator. To four decimal places, $\log 50 = 1.6990$.

Example 2   **Solve $3^{2x+1} = 12$.**

$$3^{2x+1} = 12 \qquad \text{Original equation}$$

$$\log 3^{2x+1} = \log 12 \qquad \text{Property of Equality for Logarithmic Functions.}$$

$$(2x+1)\log 3 = \log 12 \qquad \text{Power Property of Logarithms}$$

$$2x + 1 = \frac{\log 12}{\log 3} \qquad \text{Divide each side by log 3.}$$

$$2x = \frac{\log 12}{\log 3} - 1 \qquad \text{Subtract 1 from each side.}$$

$$x = \frac{1}{2}\left(\frac{\log 12}{\log 3} - 1\right) \qquad \text{Multiply each side by } \tfrac{1}{2}.$$

$$x = \frac{1}{2}\left(\frac{1.0792}{0.4771} - 1\right) \qquad \text{Use a calculator.}$$

$$x \approx 0.6309$$

Exercises

Use a calculator to evaluate each expression to the nearest ten-thousandth.

1. $\log 18$ **2.** $\log 39$ **3.** $\log 120$

4. $\log 5.8$ **5.** $\log 42.3$ **6.** $\log 0.003$

Solve each equation or inequality. Round to the nearest ten-thousandth.

7. $4^{3x} = 12$ **8.** $6^{x+2} = 18$

9. $5^{4x-2} = 120$ **10.** $7^{3x-1} \geq 21$

11. $2.4^{x+4} = 30$ **12.** $6.5^{2x} \geq 200$

13. $3.6^{4x-1} = 85.4$ **14.** $2^{x+5} = 3^{x-2}$

15. $9^{3x} = 4^{5x+2}$ **16.** $6^{x-5} = 2^{7x+3}$

8-6 Study Guide and Intervention (continued)

Common Logarithms

Change of Base Formula The following formula is used to change expressions with different logarithmic bases to common logarithm expressions.

Change of Base Formula	For all positive numbers a, b, and n, where $a \neq 1$ and $b \neq 1$, $\log_a n = \dfrac{\log_b n}{\log_b a}$

Example Express $\log_8 15$ in terms of common logarithms. Then round to the nearest ten-thousandth.

$\log_8 15 = \dfrac{\log_{10} 15}{\log_{10} 8}$ Change of Base Formula

≈ 1.3023 Simplify.

The value of $\log_8 15$ is approximately 1.3023.

Exercises

Express each logarithm in terms of common logarithms. Then approximate its value to the nearest ten-thousandth.

1. $\log_3 16$ **2.** $\log_2 40$ **3.** $\log_5 35$

4. $\log_4 22$ **5.** $\log_{12} 200$ **6.** $\log_2 50$

7. $\log_5 0.4$ **8.** $\log_3 2$ **9.** $\log_4 28.5$

10. $\log_3 (20)^2$ **11.** $\log_6 (5)^4$ **12.** $\log_8 (4)^5$

13. $\log_5 (8)^3$ **14.** $\log_2 (3.6)^6$ **15.** $\log_{12} (10.5)^4$

16. $\log_3 \sqrt{150}$ **17.** $\log_4 \sqrt[3]{39}$ **18.** $\log_5 \sqrt[4]{1600}$

8-7 Study Guide and Intervention

Base e and Natural Logarithms

Base e and Natural Logarithms The irrational number $e \approx 2.71828...$ often occurs as the base for exponential and logarithmic functions that describe real-world phenomena.

Natural Base e	As n increases, $\left(1 + \frac{1}{n}\right)^n$ approaches $e \approx 2.71828....$ $\ln x = \log_e x$

The functions $f(x) = e^x$ and $f(x) = \ln x$ are inverse functions.

Inverse Property of Base e and Natural Logarithms	$e^{\ln x} = x$ $\ln e^x = x$

Natural base expressions can be evaluated using the e^x and ln keys on your calculator.

Example 1   Write a logarithmic equation equivalent to $e^{2x} = 7$.

$e^{2x} = 7 \rightarrow \log_e 7 = 2x$

$\qquad\qquad 2x = \ln 7$

Example 2   Write each logarithmic equation in exponential form.

a. $\ln x \approx 0.3345$

$\ln x \approx 0.3345 \rightarrow \log_e x \approx 0.3345$

$\qquad\qquad\qquad x \approx e^{0.3345}$

b. $\ln 42 = x$

$\ln 42 = x \rightarrow \log_e 42 = x$

$\qquad\qquad\qquad 42 = e^x$

Exercises

Write an equivalent exponential or logarithmic equation.

1. $e^{15} = x$ **2.** $e^{3x} = 45$ **3.** $\ln 20 = x$ **4.** $\ln x = 8$

5. $e^{-5x} = 0.2$ **6.** $\ln (4x) = 9.6$ **7.** $e^{8.2} = 10x$ **8.** $\ln 0.0002 = x$

Evaluate each logarithm to the nearest ten–thousandth.

9. $\ln 12,492$ **10.** $\ln 50.69$ **11.** $\ln 9275$ **12.** $\ln 0.835$

13. $\ln 943 - \ln 181$ **14.** $\ln 67 + \ln 103$ **15.** $\ln 931 \cdot \ln 32$ **16.** $\ln (139 - 45)$

8-7 Study Guide and Intervention (continued)

Base e and Natural Logarithms

Equations and Inequalities with e and ln All properties of logarithms from earlier lessons can be used to solve equations and inequalities with natural logarithms.

Example Solve each equation or inequality.

a. $3e^{2x} + 2 = 10$

$3e^{2x} + 2 = 10$	Original equation
$3e^{2x} = 8$	Subtract 2 from each side.
$e^{2x} = \dfrac{8}{3}$	Divide each side by 3.
$\ln e^{2x} = \ln \dfrac{8}{3}$	Property of Equality for Logarithms
$2x = \ln \dfrac{8}{3}$	Inverse Property of Exponents and Logarithms
$x = \dfrac{1}{2} \ln \dfrac{8}{3}$	Multiply each side by $\frac{1}{2}$.
$x \approx 0.4904$	Use a calculator.

b. $\ln (4x - 1) < 2$

$\ln (4x - 1) < 2$	Original inequality
$e^{\ln (4x - 1)} < e^2$	Write each side using exponents and base e.
$0 < 4x - 1 < e^2$	Inverse Property of Exponents and Logarithms
$1 < 4x < e^2 + 1$	Addition Property of Inequalities
$\dfrac{1}{4} < x < \dfrac{1}{4}(e^2 + 1)$	Multiplication Property of Inequalities
$0.25 < x < 2.0973$	Use a calculator.

Exercises

Solve each equation or inequality. Round to the nearest ten-thousandth.

1. $e^{4x} = 120$

2. $e^x \leq 25$

3. $e^{x-2} + 4 = 21$

4. $\ln 6x \geq 4$

5. $\ln (x + 3) - 5 = -2$

6. $e^{-8x} \leq 50$

7. $e^{4x-1} - 3 = 12$

8. $\ln (5x + 3) = 3.6$

9. $2e^{3x} + 5 = 2$

10. $6 + 3e^{x+1} = 21$

11. $\ln (2x - 5) = 8$

12. $\ln 5x + \ln 3x > 9$

8-8　Study Guide and Intervention

Using Exponential and Logarithmic Functions

Exponential Growth and Decay

Exponential Growth	$f(x) = ae^{kt}$ where a is the initial value of y, t is time in years, and k is a constant representing the **rate of continuous growth**.
Exponential Decay	$f(x) = ae^{-kt}$ where a is the initial value of y, t is time in years, and k is a constant representing **the rate of continuous decay**.

Example　POPULATION In 2000, the world population was estimated to be 6.124 billion people. In 2005, it was 6.515 billion.

a. Determine the value of k, the world's relative rate of growth

$$y = ae^{kt}$$ Formula for continuous growth.

$$6.515 = 6.124e^{k(5)}$$ $y = 6.515$, $a = 6.124$, and $t = 2005 - 2000$ or 5

$$\frac{6.515}{6.124} = e^{5k}$$ Divide each side by 6.124.

$$\ln \frac{6.515}{6.124} = \ln e^{5k}$$ Property of Equality for Logarithmic Functions

$$\ln \frac{6.515}{6.124} = 5k$$ $\ln e^x = x$

$$0.01238 = k$$ Divide each side by 5 and use a calculator.

The world's relative rate of growth is about 0.01238 or 1.2%

b. When will the world's population reach 7.5 billion people?

$$7.5 = 6.124e^{0.01238t}$$ $y = 7.5$, $a = 6.124$, and $k = 0.01238$

$$\frac{7.5}{6.124} = e^{0.01238t}$$ Divide each side by 6.124.

$$\ln \frac{7.5}{6.124} = e^{0.01238t}$$ Property of Equality for Logarithmic Functions

$$\ln \frac{7.5}{6.124} = 0.01238t$$ $\ln e^x = x$

$$16.3722 = t$$ Divide each side by 0.01238 and use a calculator.

The world's population will reach 7.5 billion in 2016.

Exercises

1. **CARBON DATING** Use the formula $y = ae^{-0.00012t}$, where a is the initial amount of carbon 14, t is the number of years ago the animal lived, and y is the remaining amount after t years.

    a. How old is a fossil that has lost 95% of its Carbon-14?

    b. How old is a skeleton that has 95% of its Carbon-14 remaining?

8-8 Study Guide and Intervention *(continued)*

Using Exponential and Logarithmic Functions

Logistic Growth A logistic function models the S-curve of growth of some set λ. The initial stage of growth is approximately exponential; then, as saturation begins, the growth slows, and at some point, growth stops.

Example **The population of a certain species of fish in a lake after t years is given by $P(t) = \dfrac{1880}{1 + 1.42e^{-0.037t}}$.**

a. Graph the function.

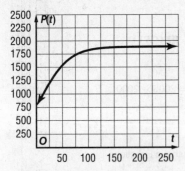

b. Find the horizontal asymptote. What does it represent in the situation?

The horizontal asymptote is $P(t) = 1880$. The population of fish will reach a ceiling of 1880.

c. When will the population reach 1875?

$$1875 = \frac{1880}{1 + 1.42e^{-0.037t}}$$ Replace $P(t)$ with 1875.

$$1875(1 + 1.42e^{-0.037t}) = 1880$$ Multiply each side by $(1 + 1.42e^{-0.037t})$.

$$2662.5e^{-0.037t} = 5$$ Simplify and subtract 1875 from each side.

$$e^{-0.037t} = \frac{5}{2662.5}$$ Divide each side by 2662.5.

$$-0.037t = \ln \frac{5}{2662.5}$$ Take the natural logarithm of each side.

$$t = \left(\ln \frac{5}{2662.5}\right) \div (-0.037)$$ Divide each side by −0.037.

$$t \approx 169.66$$ Use a calculator.

The population will reach 1875 in about 170 years.

Exercises

1. Assume the population of gnats in a specific habitat follows the function

$$P(t) = \frac{17000}{(1 + 15e^{-0.0082t})}.$$

a. Graph the function for $0 \le t \le 500$.

b. What is the horizontal asymptote?

c. What is the maximum population?

d. When does the population reach 15,000?

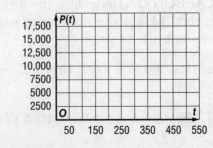

9-1 Study Guide and Intervention

Multiplying and Dividing Rational Expressions

Simplify Rational Expressions A ratio of two polynomial expressions is a **rational expression**. To simplify a rational expression, divide both the numerator and the denominator by their greatest common factor (GCF).

Multiplying Rational Expressions	For all rational expressions $\frac{a}{b}$ and $\frac{c}{d}$, $\frac{a}{b} \cdot \frac{c}{d} = \frac{ac}{bd}$, if $b \neq 0$ and $d \neq 0$.
Dividing Rational Expressions	For all rational expressions $\frac{a}{b}$ and $\frac{c}{d}$, $\frac{a}{b} \div \frac{c}{d} = \frac{ad}{bc}$, if $b \neq 0$, $c \neq 0$, and $d \neq 0$.

Example Simplify each expression.

a. $\dfrac{24a^5b^2}{(2ab)^4}$

$$\frac{24a^5b^2}{(2ab)^4} = \frac{\overset{1}{\cancel{2}} \cdot \overset{1}{\cancel{2}} \cdot \overset{1}{\cancel{2}} \cdot 3 \cdot \overset{1}{\cancel{a}} \cdot \overset{1}{\cancel{a}} \cdot \overset{1}{\cancel{a}} \cdot \overset{1}{\cancel{a}} \cdot a \cdot \overset{1}{\cancel{b}} \cdot \overset{1}{\cancel{b}}}{\underset{1}{\cancel{2}} \cdot \underset{1}{\cancel{2}} \cdot \underset{1}{\cancel{2}} \cdot 2 \cdot \underset{1}{\cancel{a}} \cdot \underset{1}{\cancel{a}} \cdot \underset{1}{\cancel{a}} \cdot \underset{1}{\cancel{a}} \cdot \underset{1}{\cancel{b}} \cdot \underset{1}{\cancel{b}} \cdot b \cdot b} = \frac{3a}{2b^2}$$

b. $\dfrac{3r^2n^3}{5t^4} \cdot \dfrac{20t^2}{9r^3n}$

$$\frac{3r^2n^3}{5t^4} \cdot \frac{20t^2}{9r^3n} = \frac{\overset{1}{\cancel{3}} \cdot \overset{1}{\cancel{r}} \cdot \overset{1}{\cancel{r}} \cdot \overset{1}{\cancel{n}} \cdot n \cdot n \cdot 2 \cdot 2 \cdot \overset{1}{\cancel{5}} \cdot \overset{1}{\cancel{t}} \cdot \overset{1}{\cancel{t}}}{\underset{1}{\cancel{5}} \cdot \underset{1}{\cancel{t}} \cdot \underset{1}{\cancel{t}} \cdot t \cdot t \cdot \underset{1}{\cancel{3}} \cdot 3 \cdot \underset{1}{\cancel{r}} \cdot \underset{1}{\cancel{r}} \cdot r \cdot \underset{1}{\cancel{n}}} = \frac{2 \cdot 2 \cdot n \cdot n}{3 \cdot r \cdot t \cdot t} = \frac{4n^2}{3rt^2}$$

c. $\dfrac{x^2 + 8x + 16}{2x - 2} \div \dfrac{x^2 + 2x - 8}{x - 1}$

$$\frac{x^2 + 8x + 16}{2x - 2} \div \frac{x^2 + 2x - 8}{x - 1} = \frac{x^2 + 8x + 16}{2x - 2} \cdot \frac{x - 1}{x^2 + 2x - 8}$$

$$= \frac{(x + 4)(x + 4)\overset{1}{\cancel{(x - 1)}}}{2\underset{1}{\cancel{(x - 1)}}(x - 2)\overset{1}{\cancel{(x + 4)}}} = \frac{x + 4}{2(x - 2)}$$

Exercises

Simplify each expression.

1. $\dfrac{(-2ab^2)^3}{20ab^4}$

2. $\dfrac{4x^2 - 12x + 9}{9 - 6x}$

3. $\dfrac{x^2 + x - 6}{x^2 - 6x - 27}$

4. $\dfrac{3m^3 - 3m}{6m^4} \cdot \dfrac{4m^5}{m + 1}$

5. $\dfrac{c^2 - 3c}{c^2 - 25} \cdot \dfrac{c^2 + 4c - 5}{c^2 - 4c + 3}$

6. $\dfrac{(m - 3)^2}{m^2 - 6m + 9} \cdot \dfrac{m^3 - 9m}{m^2 - 9}$

7. $\dfrac{6xy^4}{25z^3} \div \dfrac{18xz^2}{5y}$

8. $\dfrac{16p^2 - 8p + 1}{14p^4} \div \dfrac{4p^2 + 7p - 2}{7p^5}$

9. $\dfrac{2m - 1}{m^2 - 3m - 10} \div \dfrac{4m^2 - 1}{4m + 8}$

9-1 Study Guide and Intervention (continued)

Multiplying and Dividing Rational Expressions

Simplify Complex Fractions A **complex fraction** is a rational expression with a numerator and/or denominator that is also a rational expression. To simplify a complex fraction, first rewrite it as a division problem.

Example Simplify $\dfrac{\frac{3n-1}{n}}{\frac{3n^2+8n-3}{n^4}}$.

$$\dfrac{\frac{3n-1}{n}}{\frac{3n^2+8n-3}{n^4}} = \frac{3n-1}{n} \div \frac{3n^2+8n-3}{n^4}$$ Express as a division problem.

$$= \frac{3n-1}{n} \cdot \frac{n^4}{3n^2+8n-3}$$ Multiply by the reciprocal of the divisor.

$$= \frac{\overset{1}{\cancel{(3n-1)}}\overset{n^3}{\cancel{n^4}}}{\underset{1}{n\cancel{(3n-1)}(n+3)}}$$ Factor and eliminate.

$$= \frac{n^3}{n+3}$$ Simplify.

Exercises

Simplify each expression.

1. $\dfrac{\frac{x^3y^2z}{a^2b^2}}{\frac{a^3x^2y}{b^2}}$

2. $\dfrac{\frac{a^2bc^3}{x^2y^2}}{\frac{ab^2}{c^4x^2y}}$

3. $\dfrac{\frac{b^2-1}{3b+2}}{\frac{b+1}{3b^2-b-2}}$

4. $\dfrac{\frac{b^2-100}{b^2}}{\frac{3b^2-31b+10}{2b}}$

5. $\dfrac{\frac{x-4}{x^2+6x+9}}{\frac{x^2-2x-8}{3+x}}$

6. $\dfrac{\frac{a^2-16}{a+2}}{\frac{a^2+3a-4}{a^2+a-2}}$

7. $\dfrac{\frac{2x^2+9x+9}{x+1}}{\frac{10x^2+19x+6}{5x^2+7x+2}}$

8. $\dfrac{\frac{b+2}{b^2-6b+8}}{\frac{b^2+b-2}{b^2-16}}$

9. $\dfrac{\frac{x^2-x-2}{x^2+x-6}}{\frac{x+1}{x+3}}$

9-2 Study Guide and Intervention

Adding and Subtracting Rational Expressions

LCM of Polynomials To find the least common multiple of two or more polynomials, factor each expression. The LCM contains each factor the greatest number of times it appears as a factor.

Example 1 Find the LCM of $16p^2q^3r$, $40pq^4r^2$, and $15p^3r^4$.

$16p^2q^3r = 2^4 \cdot p^2 \cdot q^3 \cdot r$

$40pq^4r^2 = 2^3 \cdot 5 \cdot p \cdot q^4 \cdot r^2$

$15p^3r^4 = 3 \cdot 5 \cdot p^3 \cdot r^4$

LCM $= 2^4 \cdot 3 \cdot 5 \cdot p^3 \cdot q^4 \cdot r^4$

$= 240p^3q^4r^4$

Example 2 Find the LCM of $3m^2 - 3m - 6$ and $4m^2 + 12m - 40$.

$3m^2 - 3m - 6 = 3(m + 1)(m - 2)$

$4m^2 + 12m - 40 = 4(m - 2)(m + 5)$

LCM $= 12(m + 1)(m - 2)(m + 5)$

Exercises

Find the LCM of each set of polynomials.

1. $14ab^2$, $42bc^3$, $18a^2c$

2. $8cdf^3$, $28c^2f$, $35d^4f^2$

3. $65x^4y$, $10x^2y^2$, $26y^4$

4. $11mn^5$, $18m^2n^3$, $20mn^4$

5. $15a^4b$, $50a^2b^2$, $40b^8$

6. $24p^7q$, $30p^2q^2$, $45pq^3$

7. $39b^2c^2$, $52b^4c$, $12c^3$

8. $12xy^4$, $42x^2y$, $30x^2y^3$

9. $56stv^2$, $24s^2v^2$, $70t^3v^3$

10. $x^2 + 3x$, $10x^2 + 25x - 15$

11. $9x^2 - 12x + 4$, $3x^2 + 10x - 8$

12. $22x^2 + 66x - 220$, $4x^2 - 16$

13. $8x^2 - 36x - 20$, $2x^2 + 2x - 60$

14. $5x^2 - 125$, $5x^2 + 24x - 5$

15. $3x^2 - 18x + 27$, $2x^3 - 4x^2 - 6x$

16. $45x^2 - 6x - 3$, $45x^2 - 5$

17. $x^3 + 4x^2 - x - 4$, $x^2 + 2x - 3$

18. $54x^3 - 24x$, $12x^2 - 26x + 12$

9-2 **Study Guide and Intervention** *(continued)*

Adding and Subtracting Rational Expressions

Add and Subtract Rational Expressions To add or subtract rational expressions, follow these steps.

> **Step 1** Find the least common denominator (LCD). Rewrite each expression with the LCD.
> **Step 2** Add or subtract the numerators.
> **Step 3** Combine any like terms in the numerator.
> **Step 4** Factor if possible.
> **Step 5** Simplify if possible.

Example Simplify $\dfrac{6}{2x^2 + 2x - 12} - \dfrac{2}{x^2 - 4}$.

$\dfrac{6}{2x^2 + 2x - 12} - \dfrac{2}{x^2 - 4}$

$= \dfrac{6}{2(x + 3)(x - 2)} - \dfrac{2}{(x - 2)(x + 2)}$ Factor the denominators.

$= \dfrac{6(x + 2)}{2(x + 3)(x - 2)(x + 2)} - \dfrac{2 \cdot 2(x + 3)}{2(x + 3)(x - 2)(x + 2)}$ The LCD is $2(x + 3)(x - 2)(x + 2)$.

$= \dfrac{6(x + 2) - 4(x + 3)}{2(x + 3)(x - 2)(x + 2)}$ Subtract the numerators.

$= \dfrac{6x + 12 - 4x - 12}{2(x + 3)(x - 2)(x + 2)}$ Distribute.

$= \dfrac{2x}{2(x + 3)(x - 2)(x + 2)}$ Combine like terms.

$= \dfrac{x}{(x + 3)(x - 2)(x + 2)}$ Simplify.

Exercises

Simplify each expression.

1. $\dfrac{-7xy}{3x} + \dfrac{4y^2}{2y}$

2. $\dfrac{2}{x - 3} - \dfrac{1}{x - 1}$

3. $\dfrac{4a}{3bc} - \dfrac{15b}{5ac}$

4. $\dfrac{3}{x + 2} + \dfrac{4x + 5}{3x + 6}$

5. $\dfrac{3x + 3}{x^2 + 2x + 1} + \dfrac{x - 1}{x^2 - 1}$

6. $\dfrac{4}{4x^2 - 4x + 1} - \dfrac{5x}{20x^2 - 5}$

9-3 Study Guide and Intervention

Graphing Reciprocal Functions

Vertical and Horizontal Asymptotes

Parent Function of Reciprocal Functions	
Parent Function	$y = \frac{1}{x}$
Type of Graph	hyperbola
Domain	all nonzero real numbers
Range	all nonzero real numbers
Symmetry	over the x- and y-axes
Intercepts	none
Asymptotes	the x- and y-axes

Example **Identify the asymptotes, domain, and range of the function**

$f(x) = \frac{3}{x + 2}$.

Identify x values for which $f(x)$ is undefined.

$x + 2 = 0$, so $x = -2$. $f(x)$ is not defined when $x = -2$, so there is an asymptote at $x = -2$.

From $x = -2$, as x-values decrease, $f(x)$ approaches 0. As x-values increase, $f(x)$ approaches 0. So there is an asymptote at $f(x) = 0$.

The domain is all real numbers not equal to -2, and the range is all real numbers not equal to 0.

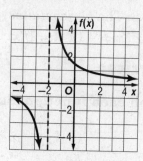

Exercises

Identify the asymptotes, domain, and range of each function.

1. $f(x) = \frac{1}{x}$

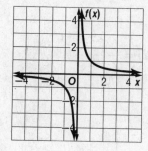

2. $f(x) = \frac{-3}{x - 1}$

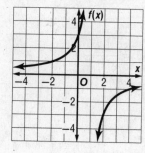

3. $f(x) = \frac{4}{x + 1} + 2$

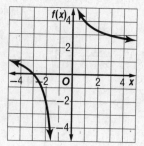

9-3 Study Guide and Intervention *(continued)*

Graphing Reciprocal Functions

Transformations of Reciprocal Functions

Equation Form	$f(x) = \dfrac{a}{x-h} + k$
Horizontal Translation	The *vertical* asymptote moves to $x = h$.
Vertical Translation	The *horizontal* asymptote moves to $y = k$.
Reflection	The graph is reflected across the *x*-axis when $a < 0$.
Compression and Expansion	The graph is compressed vertically when $|a| < 1$ and expanded vertically when $|a| > 1$.

Example

Graph $f(x) = \dfrac{\frac{-1}{2}}{x+1} - 3$. State the domain and range.

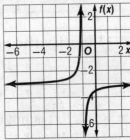

$a < 0$: The graph is reflected over the *x*-axis.

$0 < |a| < 1$: The graph is compressed vertically.

$h = -1$: The *vertical* asymptote is at $x = -1$.

$k = -3$: The *horizontal* asymptote is at $f(x) = -3$.

$D = \{x \mid x \neq -1\}$; $R = \{f(x) \mid f(x) \neq -3\}$

Exercises

Graph each function. State the domain and range.

1. $f(x) = \dfrac{1}{x} + 1$

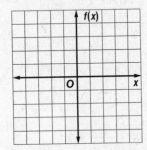

2. $f(x) = \dfrac{-2}{x-2}$

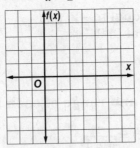

3. $f(x) = \dfrac{-1}{x-3}$

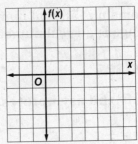

4. $f(x) = \dfrac{1}{x+5} + 3$

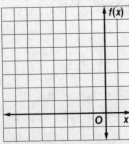

5. $f(x) = \dfrac{-2}{x-1} + 2$

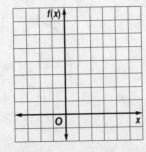

6. $f(x) = \dfrac{1}{x-3} + 4$

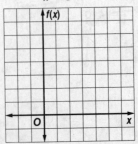

9-4 Study Guide and Intervention

Graphing Rational Functions

Vertical and Horizontal Asymptotes

Rational Function	A function with an equation of the form $f(x) = \dfrac{p(x)}{q(x)}$, where $p(x)$ and $q(x)$ are polynomial expressions and $q(x) \neq 0$.
Domain	The domain of a rational function is limited to values for which the function is defined.
Vertical Asymptote	An asymptote is a line that the graph of a function approaches. If the simplified form of the related rational expression is undefined for $x = a$, then $x = a$ is a vertical asymptote.
Horizontal Asymptote	Often a horizontal asymptote occurs in the graph of a rational function where a value is excluded from the range.

Example   Graph $f(x) = \dfrac{x^2 + x - 6}{x + 1}$.

$$\frac{x^2 + x - 6}{x + 1} = \frac{(x + 3)(x - 2)}{x + 1}$$

Therefore the graph of $f(x)$ has zeroes at $x = -3$ and $x = 2$ and a vertical asymptote at $x = -1$. Because the degree of $x^2 + x - 6$ is greater than $x + 1$, there is no horizontal asymptote. Make a table of values. Plot the points and draw the graph.

x	−5	−4	−3	−2	0	1	2	3	4
f(x)	−3.5	−2	0	4	−6	−2	0	1.5	2.8

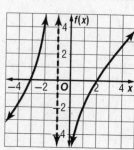

Exercises

Graph each function.

1. $f(x) = \dfrac{4}{x^2 + 3x - 10}$

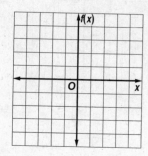

2. $f(x) = \dfrac{x^2 - 2x + 1}{x^2 + 2x + 1}$

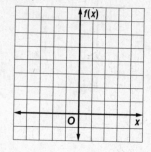

3. $f(x) = \dfrac{2x + 9}{2x^2 - x - 3}$

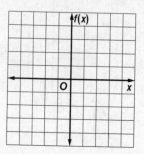

9-4   Study Guide and Intervention   *(continued)*

Graphing Rational Functions

Oblique Asymptotes and Point Discontinuity   An **oblique asymptote** is an asymptote that is neither horizontal nor vertical. In some cases, graphs of rational functions may have **point discontinuity**, which looks like a hole in the graph. That is because the function is undefined at that point.

Oblique Asymptotes	If $f(x) = \dfrac{a(x)}{b(x)}$, $a(x)$ and $b(x)$ are polynomial functions with no common factors other than 1 and $b(x) \neq 0$, then $f(x)$ has an oblique asymptote if the degree of $a(x)$ minus the degree of $b(x)$ equals 1.
Point Discontinuity	If $f(x) = \dfrac{a(x)}{b(x)}$, $b(x) \neq 0$, and $x - c$ is a factor of both $a(x)$ and $b(x)$, then there is a point discontinuity at $x = c$.

Example   Graph $f(x) = \dfrac{x - 1}{x^2 + 2x - 3}$.

$$\dfrac{x - 1}{x^2 + 2x - 3} = \dfrac{x - 1}{(x - 1)(x + 3)} \text{ or } \dfrac{1}{x + 3}$$

Therefore the graph of $f(x)$ has an asymptote at $x = -3$ and a point discontinuity at $x = 1$.

Make a table of values. Plot the points and draw the graph.

x	−2.5	−2	−1	−3.5	−4	−5
f(x)	2	1	0.5	−2	−1	−0.5

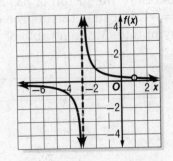

Exercises

Graph each function.

1. $f(x) = \dfrac{x^2 + 5x + 4}{x + 3}$

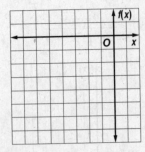

2. $f(x) = \dfrac{x^2 - x - 6}{x - 3}$

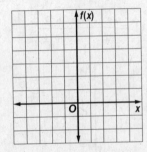

3. $f(x) = \dfrac{x^2 - 6x + 8}{x^2 - x - 2}$

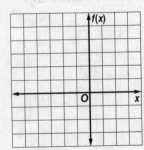

9-5 Study Guide and Intervention

Variation Functions

Direct Variation and Joint Variation

Direct Variation	y varies directly as x if there is some nonzero constant k such that $y = kx$. k is called the constant of variation.
Joint Variation	y varies jointly as x and z if there is some number k such that $y = kxz$, where $k \neq 0$.

Example 1 **If y varies directly as x and $y = 16$ when $x = 4$, find x when $y = 20$.**

$$\frac{y_1}{x_1} = \frac{y_2}{x_2} \qquad \text{Direct variation}$$

$$\frac{16}{4} = \frac{20}{x_2} \qquad y_1 = 16,\ x_1 = 4,\ \text{and } y_2 = 20$$

$$16x_2 = (20)(4) \qquad \text{Cross multiply.}$$

$$x_2 = 5 \qquad \text{Simplify.}$$

The value of x is 5 when y is 20.

Example 2 **If y varies jointly as x and z and $y = 10$ when $x = 2$ and $z = 4$, find y when $x = 4$ and $z = 3$.**

$$\frac{y_1}{x_1 z_1} = \frac{y_2}{x_2 z_2} \qquad \text{Joint variation}$$

$$\frac{10}{2 \cdot 4} = \frac{y_2}{4 \cdot 3} \qquad \begin{array}{l} y_1 = 10,\ x_1 = 2,\ z_1 = 4,\ x_2 = 4, \\ \text{and } z_2 = 3 \end{array}$$

$$120 = 8y_2 \qquad \text{Simplify.}$$

$$15 = y_2 \qquad \text{Divide each side by 8.}$$

The value of y is 15 when $x = 4$ and $z = 3$.

Exercises

1. If y varies directly as x and $y = 9$ when $x = 6$, find y when $x = 8$.

2. If y varies directly as x and $y = 16$ when $x = 36$, find y when $x = 54$.

3. If y varies directly as x and $x = 15$ when $y = 5$, find x when $y = 9$.

4. If y varies directly as x and $x = 33$ when $y = 22$, find x when $y = 32$.

5. Suppose y varies jointly as x and z. Find y when $x = 5$ and $z = 3$, if $y = 18$ when x is 3 and z is 2.

6. Suppose y varies jointly as x and z. Find y when $x = 6$ and $z = 8$, if $y = 6$ when x is 4 and z is 2.

7. Suppose y varies jointly as x and z. Find y when $x = 4$ and $z = 11$, if $y = 60$ when x is 3 and z is 5.

8. Suppose y varies jointly as x and z. Find y when $x = 5$ and $z = 2$, if $y = 84$ when x is 4 and z is 7.

9. If y varies directly as x and $y = 39$ when $x = 52$, find y when $x = 22$.

10. If y varies directly as x and $x = 60$ when $y = 75$, find x when $y = 42$.

11. Suppose y varies jointly as x and z. Find y when $x = 7$ and $z = 18$, if $y = 351$ when x is 6 and z is 13.

12. Suppose y varies jointly as x and z. Find y when $x = 5$ and $z = 27$, if $y = 480$ when x is 9 and z is 20.

9-5 Study Guide and Intervention (continued)

Variation Functions

Inverse Variation and Combined Variation

Inverse Variation	y varies inversely as x if there is some nonzero constant k such that $xy = k$ or $y = \frac{k}{x}$.
Combined Variation	y varies in combination with x and z if there is some nonzero constant k such that $yz = kx$ or $y = \frac{kx}{z}$.

Example If a varies directly as b, and a varies inversely as c, find b when a equals 10 and c equals -5, if b equals 4 when a equals -2 and c equals 3.

$a_1 = \dfrac{kb_1}{c_1}$ and $a_2 = \dfrac{kb_2}{c_2}$ Joint Variation Proportions

$k = \dfrac{a_1 c_1}{b_1}$ and $k = \dfrac{a_2 c_2}{b_2}$ Solve for k.

$\dfrac{a_1 c_1}{b_1} = \dfrac{a_2 c_2}{b_2}$ Set proportions equal to each other.

$\dfrac{(-2)3}{4} = \dfrac{10(-5)}{b_2}$ Substitute values from problem.

$(-2)3 \cdot b_2 = 10(-5)4$ Cross multiply.

$b_2 = 33\dfrac{1}{3}$ Simplify.

Exercises

1. If y varies inversely as x and $y = 12$ when $x = 10$, find y when $x = 15$.

2. If y varies inversely as x and $y = 100$ when $x = 38$, find y when $x = 76$.

3. If y varies inversely as x and $y = 32$ when $x = 42$, find y when $x = 24$.

4. If y varies inversely as x and $y = 36$ when $x = 10$, find y when $x = 30$.

5. If y varies inversely as x and $y = 18$ when $x = 124$, find y when $x = 93$.

6. If y varies inversely as x and $y = 90$ when $x = 35$, find y when $x = 50$.

7. If y varies inversely as x and $y = 42$ when $x = 48$, find y when $x = 36$.

8. If y varies inversely as x and $y = 44$ when $x = 20$, find y when $x = 55$.

9. If y varies inversely as x and $y = 80$ when $x = 14$, find y when $x = 35$.

10. If y varies inversely as x and $y = 3$ when $x = 8$, find y when $x = 40$.

11. If y varies directly as z and inversely as x and $y = 16$ and $z = 2$ when $x = 42$, find y when $x = 14$ and $z = 8$.

12. If y varies directly as z and inversely as x and $y = 23$ and $z = 1$ when $x = 12$, find y when $x = 15$ and $z = -3$.

9-6 Study Guide and Intervention

Solving Rational Equations and Inequalities

Solve Rational Equations A **rational equation** contains one or more rational expressions. To solve a rational equation, first multiply each side by the least common denominator of all of the denominators. Be sure to exclude any solution that would produce a denominator of zero.

Example Solve $\frac{9}{10} + \frac{2}{x+1} = \frac{2}{5}$. Check your solution.

$\frac{9}{10} + \frac{2}{x+1} = \frac{2}{5}$	Original equation
$10(x+1)\left(\frac{9}{10} + \frac{2}{x+1}\right) = 10(x+1)\left(\frac{2}{5}\right)$	Multiply each side by $10(x+1)$.
$9(x+1) + 2(10) = 4(x+1)$	Multiply.
$9x + 9 + 20 = 4x + 4$	Distribute.
$5x = -25$	Subtract $4x$ and 29 from each side.
$x = -5$	Divide each side by 5.

Check	$\frac{9}{10} + \frac{2}{x+1} = \frac{2}{5}$	Original equation
	$\frac{9}{10} + \frac{2}{-5+1} \overset{?}{=} \frac{2}{5}$	$x = -5$
	$\frac{18}{20} - \frac{10}{20} \overset{?}{=} \frac{2}{5}$	Simplify.
	$\frac{2}{5} = \frac{2}{5}$	

Exercises

Solve each equation. Check your solution.

1. $\frac{2y}{3} - \frac{y+3}{6} = 2$

2. $\frac{4t-3}{5} - \frac{4-2t}{3} = 1$

3. $\frac{2x+1}{3} - \frac{x-5}{4} = \frac{1}{2}$

4. $\frac{3m+2}{5m} + \frac{2m-1}{2m} = 4$

5. $\frac{4}{x-1} = \frac{x+1}{12}$

6. $\frac{x}{x-2} + \frac{4}{x-2} = 10$

7. **NAVIGATION** The current in a river is 6 miles per hour. In her motorboat Marissa can travel 12 miles upstream or 16 miles downstream in the same amount of time. What is the speed of her motorboat in still water? Is this a reasonable answer? Explain.

8. **WORK** Adam, Bethany, and Carlos own a painting company. To paint a particular house alone, Adam estimates that it would take him 4 days, Bethany estimates $5\frac{1}{2}$ days, and Carlos 6 days. If these estimates are accurate, how long should it take the three of them to paint the house if they work together? Is this a reasonable answer?

9-6 Study Guide and Intervention *(continued)*

Solving Rational Equations and Inequalities

Solve Rational Inequalities To solve a rational inequality, complete the following steps.

Step 1	State the excluded values.
Step 2	Solve the related equation.
Step 3	Use the values from steps 1 and 2 to divide the number line into regions. Test a value in each region to see which regions satisfy the original inequality.

Example Solve $\dfrac{2}{3n} + \dfrac{4}{5n} \le \dfrac{2}{3}$.

Step 1 The value of 0 is excluded since this value would result in a denominator of 0.

Step 2 Solve the related equation.

$$\dfrac{2}{3n} + \dfrac{4}{5n} = \dfrac{2}{3} \qquad \text{Related equation}$$

$$15n\left(\dfrac{2}{3n} + \dfrac{4}{5n}\right) = 15n\left(\dfrac{2}{3}\right) \qquad \text{Multiply each side by } 15n.$$

$$10 + 12 = 10n \qquad \text{Simplify.}$$

$$22 = 10n \qquad \text{Add.}$$

$$2.2 = n \qquad \text{Divide each side by 10.}$$

Step 3 Draw a number with vertical lines at the excluded value and the solution to the equation.

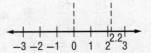

Test $n = -1$.

$-\dfrac{2}{3} + \left(-\dfrac{4}{5}\right) \le \dfrac{2}{3}$ is true.

Test $n = 1$.

$\dfrac{2}{3} + \dfrac{4}{5} \le \dfrac{2}{3}$ is *not* true.

Test $n = 3$.

$\dfrac{2}{9} + \dfrac{4}{15} \le \dfrac{2}{3}$ is true.

The solution is $n < 0$ or $n \ge 2.2$.

Exercises

Solve each inequality. Check your solutions.

1. $\dfrac{3}{a+1} \ge 3$

2. $\dfrac{1}{x} \ge 4x$

3. $\dfrac{1}{2p} + \dfrac{4}{5p} > \dfrac{2}{3}$

4. $\dfrac{3}{2x} - \dfrac{2}{x} > \dfrac{1}{4}$

5. $\dfrac{4}{x-1} + \dfrac{5}{x} < 2$

6. $\dfrac{3}{x^2-1} + 1 > \dfrac{2}{x-1}$

10-1 Study Guide and Intervention

Midpoint and Distance Formulas

The Midpoint Formula

Midpoint Formula	The midpoint M of a segment with endpoints (x_1, y_1) and (x_2, y_2) is $\left(\dfrac{x_1 + x_2}{2}, \dfrac{y_1 + y_2}{2}\right)$.

Example 1 Find the coordinates of M, the midpoint of $\overline{JK}$, for $J(4, -7)$ and $K(-2, 3)$.

$$\left(\frac{x_1 + x_2}{2}, \frac{y_1 + y_2}{2}\right) = \left(\frac{4 + (-2)}{2}, \frac{-7 + 3}{2}\right)$$
$$= \left(\frac{2}{2}, \frac{-4}{2}\right) \text{ or } (1, -2)$$

The midpoint of $\overline{JK}$ is $(1, -2)$.

Example 2 A diameter $\overline{AB}$ of a circle has endpoints $A(5, -11)$ and $B(-7, 6)$. What are the coordinates of the center of the circle?

The center of the circle is the midpoint of all of its diameters.

$$\left(\frac{x_1 + x_2}{2}, \frac{y_1 + y_2}{2}\right) = \left(\frac{5 + (-7)}{2}, \frac{-11 + 6}{2}\right)$$
$$= \left(\frac{-2}{2}, \frac{-5}{2}\right) \text{ or } \left(-1, -2\frac{1}{2}\right)$$

The circle has center $\left(-1, -2\frac{1}{2}\right)$.

Exercises

Find the midpoint of the line segment with endpoints at the given coordinates.

1. $(12, 7)$ and $(-2, 11)$ **2.** $(-8, -3)$ and $(10, 9)$ **3.** $(4, 15)$ and $(10, 1)$

4. $(-3, -3)$ and $(3, 3)$ **5.** $(15, 6)$ and $(12, 14)$ **6.** $(22, -8)$ and $(-10, 6)$

7. $(3, 5)$ and $(-6, 11)$ **8.** $(8, -15)$ and $(-7, 13)$ **9.** $(2.5, -6.1)$ and $(7.9, 13.7)$

10. $(-7, -6)$ and $(-1, 24)$ **11.** $(3, -10)$ and $(30, -20)$ **12.** $(-9, 1.7)$ and $(-11, 1.3)$

13. GEOMETRY Segment $\overline{MN}$ has midpoint P. If M has coordinates $(14, -3)$ and P has coordinates $(-8, 6)$, what are the coordinates of N?

14. GEOMETRY Circle R has a diameter $\overline{ST}$. If R has coordinates $(-4, -8)$ and S has coordinates $(1, 4)$, what are the coordinates of T?

15. GEOMETRY Segment $\overline{AD}$ has midpoint B, and $\overline{BD}$ has midpoint C. If A has coordinates $(-5, 4)$ and C has coordinates $(10, 11)$, what are the coordinates of B and D?

10-1 Study Guide and Intervention (continued)

Midpoint and Distance Formulas

The Distance Formula

Distance Formula	The distance between two points (x_1, y_1) and (x_2, y_2) is given by $d = \sqrt{(x_2 - x_1)^2 + (y_2 - y_1)^2}$.

Example 1 What is the distance between $(8, -2)$ and $(-6, -8)$?

$$d = \sqrt{(x_2 - x_1)^2 + (y_2 - y_1)^2}$$ Distance Formula

$$= \sqrt{(-6 - 8)^2 + [-8 - (-2)]^2}$$ Let $(x_1, y_1) = (8, -2)$ and $(x_2, y_2) = (-6, -8)$.

$$= \sqrt{(-14)^2 + (-6)^2}$$ Subtract.

$$= \sqrt{196 + 36} \text{ or } \sqrt{232}$$ Simplify.

The distance between the points is $\sqrt{232}$ or about 15.2 units.

Example 2 Find the perimeter and area of square $PQRS$ with vertices $P(-4, 1)$, $Q(-2, 7)$, $R(4, 5)$, and $S(2, -1)$.

Find the length of one side to find the perimeter and the area. Choose $\overline{PQ}$.

$$d = \sqrt{(x_2 - x_1)^2 + (y_2 - y_1)^2}$$ Distance Formula

$$= \sqrt{[-4 - (-2)]^2 + (1 - 7)^2}$$ Let $(x_1, y_1) = (-2, 7)$ and $(x_2, y_2) = (-4, 1)$.

$$= \sqrt{(-2)^2 + (-6)^2}$$ Subtract.

$$= \sqrt{40} \text{ or } 2\sqrt{10}$$ Simplify.

Since one side of the square is $2\sqrt{10}$, the perimeter is $8\sqrt{10}$ units. The area is $\left(2\sqrt{10}\right)^2$, or 40 units².

Exercises

Find the distance between each pair of points with the given coordinates.

1. $(3, 7)$ and $(-1, 4)$ **2.** $(-2, -10)$ and $(10, -5)$ **3.** $(6, -6)$ and $(-2, 0)$

4. $(7, 2)$ and $(4, -1)$ **5.** $(-5, -2)$ and $(3, 4)$ **6.** $(11, 5)$ and $(16, 9)$

7. $(-3, 4)$ and $(6, -11)$ **8.** $(13, 9)$ and $(11, 15)$ **9.** $(-15, -7)$ and $(2, 12)$

10. GEOMETRY Rectangle $ABCD$ has vertices $A(1, 4)$, $B(3, 1)$, $C(-3, -2)$, and $D(-5, 1)$. Find the perimeter and area of $ABCD$.

11. GEOMETRY Circle R has diameter $\overline{ST}$ with endpoints $S(4, 5)$ and $T(-2, -3)$. What are the circumference and area of the circle? (Express your answer in terms of π.)

10-2 Study Guide and Intervention

Parabolas

Equations of Parabolas A parabola is a curve consisting of all points in the coordinate plane that are the same distance from a given point (the **focus**) and a given line (the **directrix**). The following chart summarizes important information about parabolas.

Standard Form of Equation	$y = a(x - h)^2 + k$	$x = a(y - k)^2 + h$
Axis of Symmetry	$x = h$	$y = k$
Vertex	(h, k)	(h, k)
Focus	$\left(h, k + \frac{1}{4a}\right)$	$\left(h + \frac{1}{4a}, k\right)$
Directrix	$y = k - \frac{1}{4a}$	$x = h - \frac{1}{4a}$
Direction of Opening	upward if $a > 0$, downward if $a < 0$	right if $a > 0$, left if $a < 0$
Length of Latus Rectum	$\left|\frac{1}{a}\right|$ units	$\left|\frac{1}{a}\right|$ units

Example Write $y = 2x^2 - 12x - 25$ in standard form. Identify the vertex, axis of symmetry, and direction of opening of the parabola.

$y = 2x^2 - 12x - 25$	Original equation
$y = 2(x^2 - 6x) - 25$	Factor 2 from the x-terms.
$y = 2(x^2 - 6x + \blacksquare) - 25 - 2(\blacksquare)$	Complete the square on the right side.
$y = 2(x^2 - 6x + 9) - 25 - 2(9)$	The 9 added to complete the square is multiplied by 2.
$y = 2(x - 3)^2 - 43$	Write in standard form.

The vertex of this parabola is located at $(3, -43)$, the equation of the axis of symmetry is $x = 3$, and the parabola opens upward.

Exercises

Write each equation in standard form. Identify the vertex, axis of symmetry, and direction of opening of the parabola.

1. $y = x^2 + 2x + 1$

2. $y = -x^2$

3. $y = x^2 + 4x - 15$

4. $y = x^2 + 6x - 4$

5. $y = 8x - 2x^2 + 10$

6. $x = y^2 - 8y + 6$

10-2 Study Guide and Intervention *(continued)*

Parabolas

Graph Parabolas To graph an equation for a parabola, first put the given equation in standard form.

$y = a(x - h)^2 + k$ for a parabola opening up or down, or

$x = a(y - k)^2 + h$ for a parabola opening to the left or right

Use the values of a, h, and k to determine the vertex, focus, axis of symmetry, and length of the latus rectum. The vertex and the endpoints of the latus rectum give three points on the parabola. If you need more points to plot an accurate graph, substitute values for points near the vertex.

Example   **Graph** $y = \frac{1}{3}(x - 1)^2 + 2$.

In the equation, $a = \frac{1}{3}$, $h = 1$, $k = 2$.

The parabola opens up, since $a > 0$.

vertex: $(1, 2)$

axis of symmetry: $x = 1$

focus: $\left(1, 2 + \dfrac{1}{4\frac{1}{3}}\right)$ or $\left(1, 2\frac{3}{4}\right)$

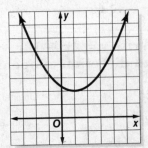

length of latus rectum: $\left|\dfrac{1}{\frac{1}{3}}\right|$ or 3 units

endpoints of latus rectum: $\left(2\frac{1}{2}, 2\frac{3}{4}\right), \left(-\frac{1}{2}, 2\frac{3}{4}\right)$

Exercises

Write an equation for each parabola described below. Then graph the equation.

1. vertex (3, 3),
    focus (3, 5)

2. vertex (4, −5),
    directrix $y = -6$

3. vertex (4, −1),
    directrix $x = 3$

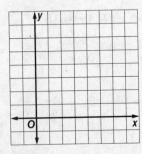

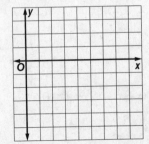

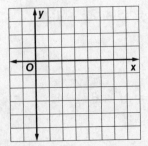

10-3 Study Guide and Intervention
Circles

Equations of Circles The equation of a circle with center (h, k) and radius r units is $(x - h)^2 + (y - k)^2 = r^2$.

A line is tangent to a circle when it touches the circle at only one point.

Example **Write an equation for a circle if the endpoints of a diameter are at $(-4, 5)$ and $(6, -3)$.**

Use the midpoint formula to find the center of the circle.

$(h, k) = \left(\dfrac{x_1 + x_2}{2}, \dfrac{y_1 + y_2}{2} \right)$ Midpoint formula

$ = \left(\dfrac{-4 + 6}{2}, \dfrac{5 + (-3)}{2} \right)$ $(x_1, y_1) = (-4, 5), (x_2, y_2) = (6, -3)$

$ = \left(\dfrac{2}{2}, \dfrac{2}{2} \right)$ or $(1, 1)$ Simplify.

Use the coordinates of the center and one endpoint of the diameter to find the radius.

$r = \sqrt{(x_2 - x_1)^2 + (y_2 - y_1)^2}$ Distance formula

$ = \sqrt{(-4 - 1)^2 + (5 - 1)^2}$ $(x_1, y_1) = (1, 1), (x_2, y_2) = (-4, 5)$

$ = \sqrt{(-5)^2 + 4^2} = \sqrt{41}$ Simplify.

The radius of the circle is $\sqrt{41}$, so $r^2 = 41$.

An equation of the circle is $(x - 1)^2 + (y - 1)^2 = 41$.

Exercises

Write an equation for the circle that satisfies each set of conditions.

1. center $(8, -3)$, radius 6

2. center $(5, -6)$, radius 4

3. center $(-5, 2)$, passes through $(-9, 6)$

4. center $(3, 6)$, tangent to the x-axis

5. center $(-4, -7)$, tangent to $x = 2$

6. center $(-2, 8)$, tangent to $y = -4$

7. center $(7, 7)$, passes through $(12, 9)$

Write an equation for each circle given the end points of a diameter.

8. $(6, 6)$ and $(10, 12)$

9. $(-4, -2)$ and $(8, 4)$

10. $(-4, 3)$ and $(6, -8)$

10-3 Study Guide and Intervention (continued)

Circles

Graph Circles To graph a circle, write the given equation in the standard form of the equation of a circle, $(x - h)^2 + (y - k)^2 = r^2$.

Plot the center (h, k) of the circle. Then use r to calculate and plot the four points $(h + r, k)$, $(h - r, k)$, $(h, k + r)$, and $(h, k - r)$, which are all points on the circle. Sketch the circle that goes through those four points.

Example Find the center and radius of the circle whose equation is $x^2 + 2x + y^2 + 4y = 11$. Then graph the circle.

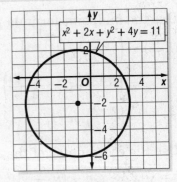

$$x^2 + 2x + y^2 + 4y = 11$$
$$x^2 + 2x + \blacksquare + y^2 + 4y + \blacksquare = 11 + \blacksquare + \blacksquare$$
$$x^2 + 2x + 1 + y^2 + 4y + 4 = 11 + 1 + 4$$
$$(x + 1)^2 + (y + 2)^2 = 16$$

Therefore, the circle has its center at $(-1, -2)$ and a radius of $\sqrt{16} = 4$. Four points on the circle are $(3, -2)$, $(-5, -2)$, $(-1, 2)$, and $(-1, -6)$.

Exercises

Find the center and radius of each circle. Then graph the circle.

1. $(x - 3)^2 + y^2 = 9$

2. $x^2 + (y + 5)^2 = 4$

3. $(x - 1)^2 + (y + 3)^2 = 9$

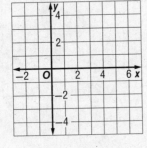
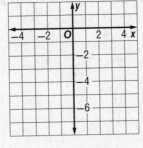
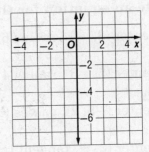

4. $(x - 2)^2 + (y + 4)^2 = 16$ **5.** $x^2 + y^2 - 10x + 8y + 16 = 0$ **6.** $x^2 + y^2 - 4x + 6y = 12$

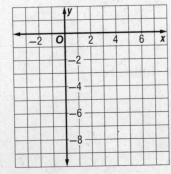
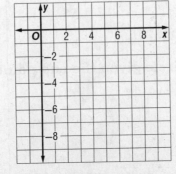
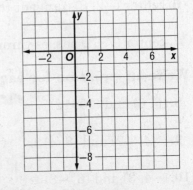

10-4   Study Guide and Intervention

Ellipses

Equations of Ellipses An **ellipse** is the set of all points in a plane such that the *sum* of the distances from two given points in the plane, called the foci, is constant. An ellipse has two axes of symmetry which contain the **major** and **minor axes**. In the table, the lengths a, b, and c are related by the formula $c^2 = a^2 - b^2$.

Standard Form of Equation	$\dfrac{(x - h)^2}{a^2} + \dfrac{(y - k)^2}{b^2} = 1$	$\dfrac{(y - k)^2}{a^2} + \dfrac{(x - h)^2}{b^2} = 1$
Center	(h, k)	(h, k)
Orientation	Horizontal	Vertical
Foci	$(h + c, k), (h - c, k)$	$(h, k - c), (h, k + c)$
Vertices	$(h + a, k), (h - a, k)$	$(h, k + a), (h, k - a)$
Length of Major Axis	$2a$ units	$2a$ units
Length of Minor Axis	$2b$ units	$2b$ units

Example   **Write an equation for the ellipse.**

The length of the major axis is the distance between $(-2, -2)$ and $(-2, 8)$. This distance is 10 units.

$2a = 10$, so $a = 5$

The foci are located at $(-2, 6)$ and $(-2, 0)$, so $c = 3$.

$b^2 = a^2 - c^2$

$ = 25 - 9$

$ = 16$

The center of the ellipse is at $(-2, 3)$, so $h = -2$, $k = 3$, $a^2 = 25$, and $b^2 = 16$. The major axis is vertical.

An equation of the ellipse is $\dfrac{(y - 3)^2}{25} + \dfrac{(x + 2)^2}{16} = 1$.

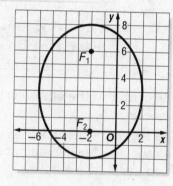

Exercises

Write an equation for an ellipse that satisfies each set of conditions.

1. vertices at $(-7, 2)$ and $(5, 2)$, co-vertices at $(-1, 0)$ and $(-1, 4)$

2. major axis 8 units long and parallel to the x-axis, minor axis 2 units long, center at $(-2, -5)$

3. vertices at $(-8, 4)$ and $(4, 4)$, foci at $(-3, 4)$ and $(-1, 4)$

4. vertices at $(3, 2)$ and $(3, -14)$, co-vertices at $(-1, -6)$ and $(7, -6)$

5. minor axis 6 units long and parallel to the x-axis, major axis 12 units long, center at $(6, 1)$

10-4 Study Guide and Intervention (continued)

Ellipses

Graph Ellipses To graph an ellipse, if necessary, write the given equation in the standard form of an equation for an ellipse.

$\dfrac{(x-h)^2}{a^2} + \dfrac{(y-k)^2}{b^2} = 1$ (for ellipse with major axis horizontal) or

$\dfrac{(y-k)^2}{a^2} + \dfrac{(x-h)^2}{b^2} = 1$ (for ellipse with major axis vertical)

Use the center (h, k) and the endpoints of the axes to plot four points of the ellipse. To make a more accurate graph, use a calculator to find some approximate values for x and y that satisfy the equation.

Example **Graph the ellipse $4x^2 + 6y^2 + 8x - 36y = -34$.**

$$4x^2 + 6y^2 + 8x - 36y = -34$$
$$4x^2 + 8x + 6y^2 - 36y = -34$$
$$4(x^2 + 2x + \blacksquare) + 6(y^2 - 6y + \blacksquare) = -34 + \blacksquare$$
$$4(x^2 + 2x + 1) + 6(y^2 - 6y + 9) = -34 + 58$$
$$4(x + 1)^2 + 6(y - 3)^2 = 24$$
$$\dfrac{(x + 1)^2}{6} + \dfrac{(y - 3)^2}{4} = 1$$

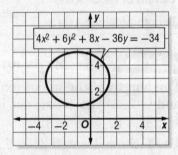

The center of the ellipse is $(-1, 3)$. Since $a^2 = 6$, $a = \sqrt{6}$.
Since $b^2 = 4$, $b = 2$.

The length of the major axis is $2\sqrt{6}$, and the length of the minor axis is 4. Since the x-term has the greater denominator, the major axis is horizontal. Plot the endpoints of the axes. Then graph the ellipse.

Exercises

Find the coordinates of the center and foci and the lengths of the major and minor axes for the ellipse with the given equation. Then graph the ellipse.

1. $\dfrac{y^2}{12} + \dfrac{x^2}{9} = 1$

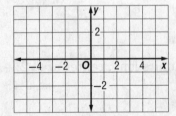

2. $\dfrac{x^2}{25} + \dfrac{y^2}{4} = 1$

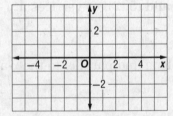

3. $x^2 + 4y^2 + 24y = -32$

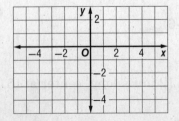

4. $9x^2 + 6y^2 - 36x + 12y = 12$

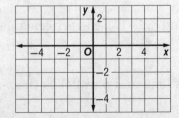

10-5 Study Guide and Intervention

Hyperbolas

Equations of Hyperbolas A **hyperbola** is the set of all points in a plane such that the absolute value of the *difference* of the distances from any point on the hyperbola to any two given points in the plane, called the **foci**, is constant.

In the table, the lengths a, b, and c are related by the formula $c^2 = a^2 + b^2$.

Standard Form	$\dfrac{x^2}{a^2} - \dfrac{y^2}{b^2} = 1$	$\dfrac{y^2}{a^2} - \dfrac{x^2}{b^2} = 1$
Orientation	Horizontal	Vertical
Foci	$(\pm c, 0)$	$(0, \pm c)$
Length of Transverse Axis	$2a$ units	$2a$ units
Length of Conjugate Axis	$2b$ units	$2b$ units
Equations of Asymptotes	$y = \pm\dfrac{b}{a}x$	$y = \pm\dfrac{a}{b}x$

Example Write an equation for the hyperbola with vertices $(0, 4)$ and $(0, -4)$ and the equation of the asymptotes is $y = \pm\dfrac{4}{3}x$.

Step 1 Find the center.
The vertices are equidistant from the center.
The center of the hyperbola is at $(0, 0)$.

Step 2 Find the values of a and b.
The hyperbola is vertical, so $a = 4$.
From the asymptotes, $b = 3$.
The value of c is not needed.

Step 3 Write the equation.
The equation for the hyperbola is $\dfrac{y^2}{16} - \dfrac{x^2}{9} = 1$.

Exercises

Write an equation for each hyperbola.

1. vertices $(-7, 0)$ and $(7, 0)$, conjugate axis of length 10

2. vertices $(0, 3)$ and $(0, -3)$, equation of asymptotes $y = \pm\dfrac{3}{8}x$

3. vertices $(0, 5)$ and $(0, -5)$, conjugate axis of length 4

4. vertices $(-8, 0)$ and $(8, 0)$, equation of asymptotes $y = \pm\dfrac{1}{6}x$

5. vertices $(-9, 0)$ and $(9, 0)$, conjugate axis of length 8

10-5 Study Guide and Intervention (continued)

Hyperbolas

Graph Hyperbolas To graph a hyperbola, write the given equation in the standard form of an equation for a hyperbola.

$$\frac{(x-h)^2}{a^2} - \frac{(y-k)^2}{b^2} = 1 \text{ if the branches of the hyperbola open left and right, or}$$

$$\frac{(y-k)^2}{a^2} - \frac{(x-h)^2}{b^2} = 1 \text{ if the branches of the hyperbola open up and down}$$

Graph the point (h, k), which is the center of the hyperbola. Draw a rectangle with dimensions $2a$ and $2b$ and center (h, k). If the hyperbola opens left and right, the vertices are $(h - a, k)$ and $(h + a, k)$. If the hyperbola opens up and down, the vertices are $(h, k - a)$ and $(h, k + a)$.

Example **Draw the graph of $6y^2 - 4x^2 - 36y - 8x = -26$.**

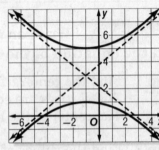

Complete the squares to get the equation in standard form.

$6y^2 - 4x^2 - 36y - 8x = -26$

$6(y^2 - 6y + \blacksquare) - 4(x^2 + 2x + \blacksquare) = -26 + \blacksquare$

$6(y^2 - 6y + 9) - 4(x^2 + 2x + 1) = -26 + 50$

$6(y - 3)^2 - 4(x + 1)^2 = 24$

$\frac{(y-3)^2}{4} - \frac{(x+1)^2}{6} = 1$

The center of the hyperbola is $(-1, 3)$.

According to the equation, $a^2 = 4$ and $b^2 = 6$, so $a = 2$ and $b = \sqrt{6}$.

The transverse axis is vertical, so the vertices are $(-1, 5)$ and $(-1, 1)$. Draw a rectangle with vertical dimension 4 and horizontal dimension $2\sqrt{6} \approx 4.9$. The diagonals of this rectangle are the asymptotes. The branches of the hyperbola open up and down. Use the vertices and the asymptotes to sketch the hyperbola.

Exercises

Graph each hyperbola. Identify the vertices, foci, and asymptotes.

1. $\frac{x^2}{4} - \frac{y^2}{16} = 1$ **2.** $(y - 3)^2 - \frac{(x+2)^2}{9} = 1$ **3.** $\frac{y^2}{16} - \frac{x^2}{9} = 1$

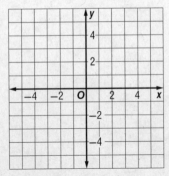
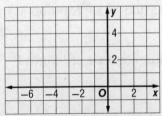
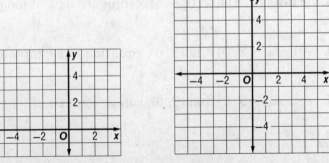

10-6 Study Guide and Intervention

Identifying Conic Sections

Conics in Standard Form Any conic section in the coordinate plane can be described by an equation of the form

$$Ax^2 + Bxy + Cy^2 + Dx + Ey + F = 0, \text{ where } A, B, \text{ and } C \text{ are not all zero.}$$

One way to tell what kind of conic section an equation represents is to rearrange terms and complete the square, if necessary, to get one of the standard forms from an earlier lesson. This method is especially useful if you are going to graph the equation.

Example Write the equation $3x^2 - 4y^2 - 30x - 8y + 59 = 0$ in standard form. State whether the graph of the equation is a *parabola, circle, ellipse,* or *hyperbola*.

$3x^2 - 4y^2 - 30x - 8y + 59 = 0$	Original equation
$3x^2 - 30x - 4y^2 - 8y = -59$	Isolate terms.
$3(x^2 - 10x + \blacksquare) - 4(y^2 + 2y + \blacksquare) = -59 + \blacksquare + \blacksquare$	Factor out common multiples.
$3(x^2 - 10x + 25) - 4(y^2 + 2y + 1) = -59 + 3(25) + (-4)(1)$	Complete the squares.
$3(x - 5)^2 - 4(y + 1)^2 = 12$	Simplify.
$\dfrac{(x - 5)^2}{4} - \dfrac{(y + 1)^2}{3} = 1$	Divide each side by 12.

The graph of the equation is a hyperbola with its center at $(5, -1)$. The length of the transverse axis is 4 units and the length of the conjugate axis is $2\sqrt{3}$ units.

Exercises

Write each equation in standard form. State whether the graph of the equation is a *parabola, circle, ellipse,* or *hyperbola*. Then graph the equation.

1. $4x^2 + 48x + y + 158 = 0$

2. $3x^2 + y^2 - 48x - 4y + 184 = 0$

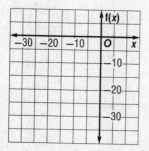

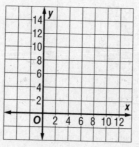

3. $-3x^2 + 2y^2 - 18x + 20y + 5 = 0$

4. $x^2 + y^2 + 8x + 2y + 8 = 0$

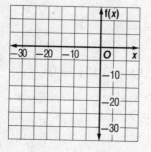

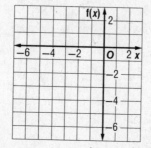

10-6 Study Guide and Intervention *(continued)*

Identifying Conic Sections

Identify Conic Sections If you are given an equation of the form

$$Ax^2 + Bxy + Cy^2 + Dx + Ey + F = 0,$$

you can determine the type of conic section just by considering the following chart.

Discriminant	Type of Conic Section
$B^2 - 4AC = 0$	parabola
$B = 0$ and $A = C$	circle
$B^2 - 4AC < 0$; either $B \neq 0$ or $A \neq C$	ellipse
$B^2 - 4AC > 0$	hyperbola

Example **Without writing the equation in standard form, state whether the graph of each equation is a *parabola*, *circle*, *ellipse*, or *hyperbola*.**

a. $3x^2 - 3y^2 - 5x + 12 = 0$
$B^2 - 4AC = 0^2 - 4(3)(-3)$
$= 36$
The graph of the equation is a hyperbola.

b. $y^2 = 7y - 2x + 13$
$B^2 - 4AC = 0^2 - 4(0)(1)$
$= 0$
The graph of the equation is a parabola.

Exercises

Without writing the equation in standard form, state whether the graph of each equation is a *parabola*, *circle*, *ellipse*, or *hyperbola*.

1. $x^2 = 17x - 5y + 8$

2. $2x^2 + 2y^2 - 3x + 4y = 5$

3. $4x^2 - 8x = 4y^2 - 6y + 10$

4. $8(2xy + x - x^2) = 4(2y^2 - y) - 100$

5. $6y^2 + 4xy - 18 = 24 - 4x^2$

6. $y = 27x - y^2$

7. $x^2 = 4(y - y^2) + 2x - 1$

8. $10x + 1xy - x^2 - 2y^2 = 5y$

9. $x = y^2 - 5y + x^2 - 5$

10. $11x^2 - 7y^2 = 77$

11. $3x^2 + 4y^2 = 50 + y^2$

12. $y^2 = 8x - 11$

13. $9y^2 - 99y = 3(3x - 3x^2)$

14. $6x^2 - 9xy - 4 = 5y^2 - 3$

15. $111 = 11x^2 + 10y^2$

16. $120x^2 - 119y^2 + 118x - 117y = 0$

17. $3x^2 = 4y^2 - xy + 12$

18. $150 - x^2 = 120 - y$

10-7 Study Guide and Intervention

Solving Linear-Nonlinear Systems

Systems of Equations Like systems of linear equations, systems of linear-nonlinear equations can be solved by substitution and elimination. If the graphs are a conic section and a line, the system will have 0, 1, or 2 solutions. If the graphs are two conic sections, the system will have 0, 1, 2, 3, or 4 solutions.

Example Solve the system of equations. $y = x^2 - 2x - 15$
$x + y = -3$

Rewrite the second equation as $y = -x - 3$ and substitute it into the first equation.

$-x - 3 = x^2 - 2x - 15$

$0 = x^2 - x - 12$ Add $x + 3$ to each side.

$0 = (x - 4)(x + 3)$ Factor.

Use the Zero Product property to get

$x = 4$ or $x = -3$.

Substitute these values for x in $x + y = -3$:

$4 + y = -3$ or $-3 + y = -3$

$y = -7$ $y = 0$

The solutions are $(4, -7)$ and $(-3, 0)$.

Exercises

Solve each system of equations.

1. $y = x^2 - 5$
$y = x - 3$

2. $x^2 + (y - 5)^2 = 25$
$y = -x^2$

3. $x^2 + (y - 5)^2 = 25$
$y = x^2$

4. $x^2 + y^2 = 9$
$x^2 + y = 3$

5. $x^2 - y^2 = 1$
$x^2 + y^2 = 16$

6. $y = x - 3$
$x = y^2 - 4$

10-7 Study Guide and Intervention (continued)

Solving Linear-Nonlinear Systems

Systems of Inequalities Systems of linear-nonlinear inequalities can be solved by graphing.

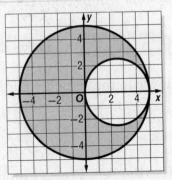

Example 1 **Solve the system of inequalities by graphing.**

$x^2 + y^2 \leq 25$

$\left(x - \dfrac{5}{2}\right)^2 + y^2 \geq \dfrac{25}{4}$

The graph of $x^2 + y^2 \leq 25$ consists of all points on or inside the circle with center $(0, 0)$ and radius 5. The graph of

$\left(x - \dfrac{5}{2}\right)^2 + y^2 \geq \dfrac{25}{4}$ consists of all points on or outside the

circle with center $\left(\dfrac{5}{2}, 0\right)$ and radius $\dfrac{5}{2}$. The solution of the

system is the set of points in both regions.

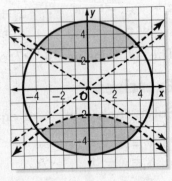

Example 2 **Solve the system of inequalities by graphing.**

$x^2 + y^2 \leq 25$

$\dfrac{y^2}{4} - \dfrac{x^2}{9} > 1$

The graph of $x^2 + y^2 \leq 25$ consists of all points on or inside the circle with center $(0, 0)$ and radius 5. The graph of

$\dfrac{y^2}{4} - \dfrac{x^2}{9} > 1$ are the points "inside" but not on the branches of

the hyperbola shown. The solution of the system is the set of points in both regions.

Exercises

Solve each system of inequalities by graphing.

1. $\dfrac{x^2}{16} + \dfrac{y^2}{4} \leq 1$

$y > \dfrac{1}{2}x - 2$

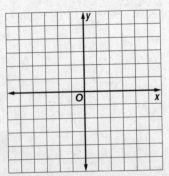

2. $x^2 + y^2 \leq 169$

$x^2 + 9y^2 \geq 225$

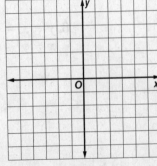

3. $y \geq (x - 2)^2$

$(x + 1)^2 + (y + 1)^2 \leq 16$

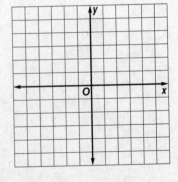

11-1 Study Guide and Intervention

Sequences as Functions

Arithmetic Sequences An **arithmetic sequence** is a sequence of numbers in which each **term** is found by adding the **common difference** to the preceding term.

nth Term of an Arithmetic Sequence	$a_n = a_1 + (n - 1)d$, where a_1 is the first term, d is the common difference, and n is any positive integer

Example **Find the next four terms of the arithmetic sequence 7, 11, 15, ... Then graph the first seven terms of the sequence.**

Find the common difference by subtracting two consecutive terms.

$11 - 7 = 4$ and $15 - 11 = 4$, so $d = 4$.

Now add 4 to the third term of the sequence, and then continue adding 4 until the four terms are found. The next four terms of the sequence are 19, 23, 27, and 31.

Plot each point $(1, 7)$, $(2, 11)$, $(3, 15)$, $(4, 19)$, $(5, 23)$, $(6, 27)$, and $(7, 31)$ on a graph.

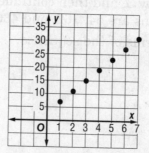

Exercises

Find the next four terms of each arithmetic sequence. Then graph the sequence.

1. 106, 111, 116, ...

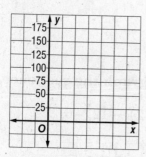

2. −28, −31, −34, ...

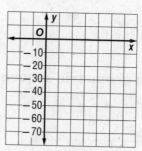

3. 207, 194, 181, ...

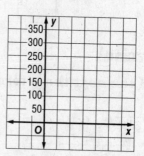

4. −30, −20, −10, ...

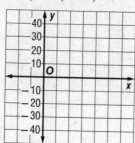

5. 13, 7, 1, ...

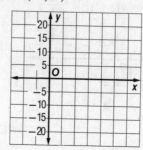

6. 151, 177, 203, ...

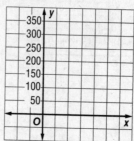

11-1 Study Guide and Intervention *(continued)*

Sequences as Functions

Geometric Sequences

Term	Definition	Example
Common Ratio	$r = a_{n+1} \div a_n$	The common ratio in a geometric sequence with consecutive terms ...5, 10 ... is $10 \div 5 = 2$.
*n*th Term of a Geometric Sequence	$a_n = a_1 \, (r^{n-1})$ where a_1 is the first term and r is the common ratio.	The fourth term of the geometric sequence with first term 5 and common ratio 2 is $5 \, (2^{4-1}) = 40$.

Example **Find the next three terms of the geometric sequence 2, 6, 18 ... .**
Then graph the sequence.

Find the common ratio by dividing two consecutive terms.

$6 \div 2 = 3$ and $18 \div 6 = 3$, so $r = 3$.

Now multiply the third term of the sequence by 3, and then continue multiplying by 3 until the three terms are found. The next three terms are 54, 162, and 486.

Find the domain and range for the first six terms of the sequence.
Domain: {1, 2, 3, 4, 5, 6}
Range: {2, 6, 18, 54, 162, 486}

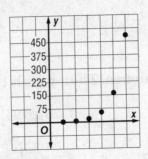

Exercises

Find the next three terms of each geometric sequence. Then graph the sequence.

1. $\frac{1}{16}, \frac{1}{4}, 1, \ldots$

2. $20, 4, \frac{4}{5}, \ldots$

3. $-24, -12, -6, \ldots$

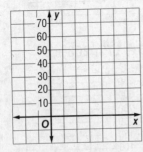

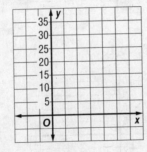

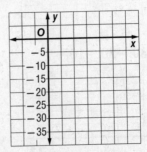

11-2 Study Guide and Intervention

Arithmetic Sequences and Series

Arithmetic Sequences

Term	Definition	Example
Common Difference	$d = a_{n+1} - a_n$	The common difference in an arithmetic sequence with consecutive terms ... 5, 7, ... is $7 - 5 = 2$.
nth Term of an Arithmetic Sequence	$a_n = a_1 + (n - 1) d$ where a_1 is the common difference and n is any positive integer.	The fifth term of the arithmetic sequence with first term 3 and common difference 2 is $3 + (4 \times 2) = 11$.

Example 1 **Find the thirteenth term of the arithmetic sequence with $a_1 = 21$ and $d = -6$.**

Use the formula for the nth term of an arithmetic sequence with $a_1 = 21$, $n = 13$, and $d = -6$.

$a_n = a_1 + (n - 1) d$ Formula for the nth term

$a_{13} = 21 + (13 - 1)(-6)$ $n = 13, a_1 = 21, d = -6$

$a_{13} = -51$

Example 2 **Write an equation for the nth term of the arithmetic sequence $-14, -5, 4, 13, \ldots$.**

In this sequence, $a_1 = -14$ and $d = 9$.   Use the formula for a_n to write an equation.

$a_n = a_1 + (n - 1) d$ Formula for the nth term

$a_n = -14 + (n - 1)(9)$ $a_1 = -14, d = 9$

$a_n = -14 + 9n - 9$ Distributive Property

$a_n = 9n - 23$ Simplify.

Exercises

Find the indicated term of each arithmetic sequence.

1. Find the twentieth term of the arithmetic sequence with $a_1 = 15$ and $d = 4$.

2. Find the seventh term of the arithmetic sequence with $a_1 = -81$ and $d = 12$.

3. Find the eleventh term of the arithmetic sequence with $a_1 = 42$ and $d = -5$.

4. Find a_{31} of the arithmetic sequence 18, 15, 12, 9, ...

5. Find a_{100} of the arithmetic sequence $-63, -58, -53, -48, \ldots$.

Write an equation for the nth term of each arithmetic sequence.

6. $a_1 = 15$ and $d = 38$

7. $a_1 = 72$ and $d = -13$

8. $-56, -39, -22, -5, \ldots$

9. $-94, -52, -10, 32, \ldots$

10. 63, 70, 77, 84, ...

11-2 Study Guide and Intervention (continued)

Arithmetic Sequences and Series

Arithmetic Series A shorthand notation for representing a series makes use of the Greek letter Σ. The **sigma notation** for the series $6 + 12 + 18 + 24 + 30$ is $\sum\limits_{n=1}^{5} 6n$.

Partial Sum of an Arithmetic Series	The sum S_n of the first n terms of an arithmetic series is given by the formula $S_n = \frac{n}{2}[2a_1 + (n-1)d]$ or $S_n = \frac{n}{2}(a_1 + a_n)$

Example 1 Find S_n for the arithmetic series with $a_1 = 14$, $a_n = 101$, and $n = 30$.

Use the sum formula for an arithmetic series.

$S_n = \frac{n}{2}(a_1 + a_n)$ Sum formula

$S_{30} = \frac{30}{2}(14 + 101)$ $n = 30, a_1 = 14, a_n = 101$

     $= 15(115)$ Simplify.

     $= 1725$ Multiply.

The sum of the series is 1725.

Example 2 Evaluate $\sum\limits_{k=1}^{18} (3k + 4)$.

The sum is an arithmetic series with common difference 3. Substituting $k = 1$ and $k = 18$ into the expression $3k + 4$ gives $a_1 = 3(1) + 4 = 7$ and $a_{18} = 3(18) + 4 = 58$. There are 18 terms in the series, so $n = 18$. Use the formula for the sum of an arithmetic series.

$S_n = \frac{n}{2}(a_1 + a_n)$ Sum formula

$S_{18} = \frac{18}{2}(7 + 58)$ $n = 18, a_1 = 7, a_n = 58$

     $= 9(65)$ Simplify.

     $= 585$ Multiply.

So $\sum\limits_{k=1}^{18} (3k + 4) = 585$.

Exercises

Find the sum of each arithmetic series.

1. $a_1 = 12, a_n = 100,$
$n = 12$

2. $a_1 = 50, a_n = -50,$
$n = 15$

3. $a_1 = 60, a_n = -136,$
$n = 50$

4. $a_1 = 20, d = 4,$
$a_n = 112$

5. $a_1 = 180, d = -8,$
$a_n = 68$

6. $a_1 = -8, d = -7,$
$a_n = -71$

7. $a_1 = 42, n = 8, d = 6$

8. $a_1 = 4, n = 20, d = 2\frac{1}{2}$

9. $a_1 = 32, n = 27, d = 3$

10. $8 + 6 + 4 + \ldots + -10$

11. $16 + 22 + 28 + \ldots + 112$

12. $\sum\limits_{n=18}^{42} (4n - 9)$

13. $\sum\limits_{n=20}^{50} (3n + 4)$

14. $\sum\limits_{j=5}^{44} (7j - 3)$

11-3 Study Guide and Intervention

Geometric Sequences and Series

Geometric Sequences A **geometric sequence** is a sequence in which each term after the first is the product of the previous term and a constant called the **constant ratio**.

nth Term of a Geometric Sequence	$a_n = a_1 \cdot r^{n-1}$, where a_1 is the first term, r is the common ratio, and n is any positive integer

Example 1 **Find the next two terms of the geometric sequence 1200, 480, 192, ...**

Since $\frac{480}{1200} = 0.4$ and $\frac{190}{480} = 0.4$, the sequence has a common ratio of 0.4. The next two terms in the sequence are $192(0.4) = 76.8$ and $76.8(0.4) = 30.72$.

Example 2 **Write an equation for the nth term of the geometric sequence 3.6, 10.8, 32.4, ...**

In this sequence $a_1 = 3.6$ and $r = 3$. Use the nth term formula to write an equation.

$a_n = a_1 \cdot r^{n-1}$ Formula for nth term
$\quad = 3.6 \cdot 3^{n-1}$ $a_1 = 3.6, r = 3$

An equation for the nth term is $a_n = 3.6 \cdot 3^{n-1}$.

Exercises

Find a_n for each geometric sequence.

1. $a_1 = -10, r = 4, n = 2$

2. $a_1 = -6, r = -\frac{1}{2}, n = 8$

3. $a_3 = 9, r = -3, n = 7$

4. $a_4 = 16, r = 2, n = 10$

5. $a_4 = -54, r = -3, n = 6$

6. $a_1 = 8, r = \frac{2}{3}, n = 5$

7. $a_1 = 7, r = 3, n = 5$

8. $a_1 = 46,875, r = \frac{1}{5}, n = 7$

9. $a_1 = -34,816, r = \frac{1}{2}, n = 6$

Write an equation for the nth term of each geometric sequence.

10. 500, 350, 245, ...

11. 8, 32, 128, ...

12. 11, −24.2, 53.24, ...

13. 9, 54, 324, 1944, ...

14. 17; 187; 2057; 22,627; ...

15. −53; −424; −3392; −27,136; ...

11-3 Study Guide and Intervention *(continued)*

Geometric Sequences and Series

Geometric Series A geometric series is the sum of the terms of a geometric sequence. The sum of the first n terms of a series is denoted S_n.

Partial Sum of a Geometric Series	The sum S_n of the first n terms with the given values a_1 and n is given by the formula $S_n = \dfrac{a_1 - a_1 r^n}{1 - r}$, $r \neq 1$
	The sum S_n of the first n terms with the given values a_1 and a_n is given by the formula $S_n = \dfrac{a_1 - a_n r}{1 - r}$, $r \neq 1$

Example 1 Find $\displaystyle\sum_{k=2}^{7} 5(3)^{k-1}$.

Find a_1, r, and k. In the first term, $k = 2$ and $a_1 = 5 \cdot 3^{2-1}$ or 15. The base of the exponential function is r, so $r = 3$. There are $7 - 2 + 1$ or 6 terms, so $k = 6$.

$$S_n = \frac{a_1 - a_1 r^k}{1 - r} \qquad \text{Sum formula}$$

$$= \frac{15 - 15(3)^6}{1 - 3} \qquad a_1 = 15, r = 3, \text{ and } k = 6$$

$$= 5460 \qquad \text{Use a calculator.}$$

Example 2 Find a_1 in a geometric series for which $S_n = 1530$, $n = 8$, and $r = 2$.

$$S_n = \frac{a_1 - a_1 r^n}{1 - r} \qquad \text{Sum formula}$$

$$1530 = \frac{a_1 - a_1(2)^8}{1 - 2} \qquad S_n = 1530, r = 2, n = 8$$

$$1530 = \frac{a_1(1 - 2^8)}{1 - 2} \qquad \text{Distributive Property}$$

$$1530 = \frac{-255a_1}{-1} \qquad \text{Subtract.}$$

$$1530 = 255a_1 \qquad \text{Simplify.}$$

$$6 = a_1 \qquad \text{Divide each side by 255.}$$

Exercises

Find the sum of each geometric series.

1. $\displaystyle\sum_{k=4}^{6} 2(-3)^{k-1}$

2. $\displaystyle\sum_{k=1}^{5} (-3)(4)^{k-1}$

3. $\displaystyle\sum_{k=3}^{10} 4(-1)^{k-1}$

4. $\displaystyle\sum_{k=3}^{7} (-1)(5)^{k-1}$

5. $\displaystyle\sum_{k=5}^{15} (-10)(-1)^{k-1}$

6. $\displaystyle\sum_{k=2}^{10} 3(3)^{k-1}$

Find a_1 for each geometric series described.

7. $S_n = 720$, $n = 4$, $r = 3$

8. $S_n = 29{,}127$, $n = 9$, $r = 4$

9. $S_n = -6552$, $r = 3$, $a_n = -4374$

10. $S_n = -936$, $r = 5$, $a_n = -750$

11-4 Study Guide and Intervention

Infinite Geometric Series

Infinite Geometric Series A geometric series that does not end is called an **infinite geometric** series. Some infinite geometric series have sums, but others do not because the **partial sums** increase without approaching a limiting value.

Sum of an Infinite Geometric Series	$S = \dfrac{a_1}{1 - r}$ for $-1 < r < 1$. If $|r| \geq 1$, the infinite geometric series does not have a sum.

Example **Find the sum of each infinite series, if it exists.**

a. $75 + 15 + 3 + \ldots$

First, find the value of r to determine if the sum exists. $a_1 = 75$ and $a_2 = 15$, so $r = \dfrac{15}{75}$ or $\dfrac{1}{5}$. Since $\left|\dfrac{1}{5}\right| < 1$, the sum exists. Now use the formula for the sum of an infinite geometric series.

$S = \dfrac{a_1}{1 - r}$ Sum formula

$= \dfrac{75}{1 - \dfrac{1}{5}}$ $a_1 = 75, r = \dfrac{1}{5}$

$= \dfrac{75}{\dfrac{4}{5}}$ or 93.75 Simplify.

The sum of the series is 93.75.

b. $\displaystyle\sum_{n=1}^{\infty} 48\left(-\dfrac{1}{3}\right)^{n-1}$

In this infinite geometric series, $a_1 = 48$ and $r = -\dfrac{1}{3}$.

$S = \dfrac{a_1}{1 - r}$ Sum formula

$= \dfrac{48}{1 - \left(-\dfrac{1}{3}\right)}$ $a_1 = 48, r = -\dfrac{1}{3}$

$= \dfrac{48}{\dfrac{4}{3}}$ or 36 Simplify.

Thus $\displaystyle\sum_{n=1}^{\infty} 48\left(-\dfrac{1}{3}\right)^{n-1} = 36$.

Exercises

Find the sum of each infinite series, if it exists.

1. $a_1 = -7, r = \dfrac{5}{8}$

2. $1 + \dfrac{5}{4} + \dfrac{25}{16} + \ldots$

3. $a_1 = 4, r = \dfrac{1}{2}$

4. $\dfrac{2}{9} + \dfrac{5}{27} + \dfrac{25}{162} + \ldots$

5. $15 + 10 + 6\dfrac{2}{3} + \ldots$

6. $18 - 9 + 4\dfrac{1}{2} - 2\dfrac{1}{4} + \ldots$

7. $\dfrac{1}{10} + \dfrac{1}{20} + \dfrac{1}{40} + \ldots$

8. $1000 + 800 + 640 + \ldots$

9. $6 - 12 + 24 - 48 + \ldots$

10. $\displaystyle\sum_{n=1}^{\infty} 50\left(\dfrac{4}{5}\right)^{n-1}$

11. $\displaystyle\sum_{k=1}^{\infty} 22\left(-\dfrac{1}{2}\right)^{k-1}$

12. $\displaystyle\sum_{s=1}^{\infty} 24\left(\dfrac{7}{12}\right)^{s-1}$

11-4 Study Guide and Intervention *(continued)*

Infinite Geometric Series

Repeating Decimals A repeating decimal represents a fraction. To find the fraction, write the decimal as an infinite geometric series and use the formula for the sum.

Example **Write each repeating decimal as a fraction.**

a. $0.\overline{42}$

Write the repeating decimal as a sum.

$0.\overline{42} = 0.42424242...$

$= 42/100 + \dfrac{42}{10,000} + \dfrac{42}{1,000,000} + ...$

In this series $a_1 = \dfrac{42}{100}$ and $r = \dfrac{1}{100}$.

$S = \dfrac{a_1}{1-r}$ Sum formula

$= \dfrac{\dfrac{42}{100}}{1 - \dfrac{1}{100}}$ $a_1 = \frac{42}{100}, r = \frac{1}{100}$

$= \dfrac{\dfrac{42}{100}}{\dfrac{99}{100}}$ Subtract.

$= \dfrac{42}{99}$ or $\dfrac{14}{33}$ Simplify.

Thus $0.\overline{42} = \dfrac{14}{33}$.

b. $0.5\overline{24}$

Let $S = 0.5\overline{24}$.

$S = 0.5242424...$ Write as a repeating decimal.

$1000S = 524.242424...$ Multiply each side by 1000.

$10S = 5.242424...$ Mulitply each side by 10.

$990S = 519$ Subtract the third equation from the second equation.

$S = \dfrac{519}{990}$ or $\dfrac{173}{330}$ Simplify.

Thus, $0.5\overline{24} = \dfrac{173}{330}$.

Exercises

Write each repeating decimal as a fraction.

1. $0.\overline{2}$

2. $0.\overline{8}$

3. $0.\overline{30}$

4. $0.\overline{87}$

5. $0.\overline{10}$

6. $0.\overline{54}$

7. $0.\overline{75}$

8. $0.\overline{18}$

9. $0.\overline{62}$

10. $0.\overline{72}$

11. $0.0\overline{72}$

12. $0.0\overline{45}$

13. $0.0\overline{6}$

14. $0.0\overline{138}$

15. $0.0\overline{0138}$

16. $0.08\overline{1}$

17. $0.2\overline{45}$

18. $0.4\overline{36}$

19. $0.5\overline{4}$

20. $0.8\overline{63}$

11-5 Study Guide and Intervention
Recursion and Iteration

Special Sequences In a **recursive formula,** each succeeding term is formulated from one or more previous terms. A recursive formula for a sequence has two parts:

1. the value(s) of the first term(s), and

2. an equation that shows how to find each term from the term(s) before it.

Example Find the first five terms of the sequence in which $a_1 = 6$, $a_2 = 10$, and $a_n = 2a_{n-2}$ for $n \geq 3$.

$a_1 = 6$

$a_2 = 10$

$a_3 = 2a_1 = 2(6) = 12$

$a_4 = 2a_2 = 2(10) = 20$

$a_5 = 2a_3 = 2(12) = 24$

The first five terms of the sequence are 6, 10, 12, 20, 24.

Exercises

Find the first five terms of each sequence described.

1. $a_1 = 1$, $a_2 = 1$, $a_n = 2(a_{n-1} + a_{n-2})$, $n \geq 3$

2. $a_1 = 1$, $a_n = \dfrac{1}{1 + a_{n-1}}$, $n \geq 2$

3. $a_1 = 3$, $a_n = a_{n-1} + 2(n-2)$, $n \geq 2$

4. $a_1 = 5$, $a_n = a_{n-1} + 2$, $n \geq 2$

5. $a_1 = 0.5$, $a_n = a_{n-1} + 2n$, $n \geq 2$

6. $a_1 = 100$, $a_n = \dfrac{a_{n-1}}{n}$, $n \geq 2$

Write a recursive formula for each sequence.

7. $1, \dfrac{1}{2}, \dfrac{1}{6}, \dfrac{1}{24}, \dfrac{1}{120}, \ldots$

8. $1, -1, 2, -3, 5, -8, \ldots$

11-5 Study Guide and Intervention *(continued)*

Recursion and Iteration

Iteration Combining composition of functions with the concept of recursion leads to the process of **iteration**. Iteration is the process of composing a function with itself repeatedly.

Example **Find the first three iterates of $f(x) = 4x - 5$ for an initial value of $x_0 = 2$.**

To find the first iterate, find the value of the function for $x_0 = 2$

$x_1 = f(x_0)$ Iterate the function.

  $= f(2)$ $x_0 = 2$

  $= 4(2) - 5$ or 3 Simplify.

To find the second iteration, find the value of the function for $x_1 = 3$.

$x_2 = f(x_1)$ Iterate the function.

  $= f(3)$ $x_1 = 3$

  $= 4(3) - 5$ or 7 Simplify.

To find the third iteration, find the value of the function for $x_2 = 7$.

$x_3 = f(x_2)$ Iterate the function.

  $= f(7)$ $x_2 = 7$

  $= 4(7) - 5$ or 23 Simplify.

The first three iterates are 3, 7, and 23.

Exercises

Find the first three iterates of each function for the given initial value.

1. $f(x) = x - 1; x_0 = 4$ **2.** $f(x) = x^2 - 3x; x_0 = 1$ **3.** $f(x) = x^2 + 2x + 1; x_0 = -2$

4. $f(x) = 4x - 6; x_0 = -5$ **5.** $f(x) = 6x - 2; x_0 = 3$ **6.** $f(x) = 100 - 4x; x_0 = -5$

7. $f(x) = 3x - 1; x_0 = 47$ **8.** $f(x) = x^3 - 5x^2; x_0 = 1$ **9.** $f(x) = 10x - 25; x_0 = 2$

10. $f(x) = 4x^2 - 9; x_0 = -1$ **11.** $f(x) = 2x^2 + 5; x_0 = -4$ **12.** $f(x) = \dfrac{x - 1}{x + 2}; x_0 = 1$

13. $f(x) = \dfrac{1}{2}(x + 11); x_0 = 3$ **14.** $f(x) = \dfrac{3}{x}; x_0 = 9$ **15.** $f(x) = x - 4x^2; x_0 = 1$

16. $f(x) = x + \dfrac{1}{x}; x_0 = 2$ **17.** $f(x) = x^3 - 5x^2 + 8x - 10;$ $x_0 = 1$ **18.** $f(x) = x^3 - x^2; x_0 = -2$

11-6 Study Guide and Intervention

The Binomial Theorem

Pascal's Triangle **Pascal's triangle** is the pattern of coefficients of powers of binomials displayed in triangular form. Each row begins and ends with 1 and each coefficient is the sum of the two coefficients above it in the previous row.

Pascal's Triangle		
	$(a + b)^0$	1
	$(a + b)^1$	1 1
	$(a + b)^2$	1 2 1
	$(a + b)^3$	1 3 3 1
	$(a + b)^4$	1 4 6 4 1
	$(a + b)^5$	1 5 10 10 5 1

Example Use Pascal's triangle to find the number of possible sequences consisting of 3 as and 2 bs.

The coefficient 10 of the a^3b^2-term in the expansion of $(a + b)^5$ gives the number of sequences that result in three as and two bs.

Exercises

Expand each binomial.

1. $(a + 5)^4$

2. $(x - 2y)^6$

3. $(j - 3k)^5$

4. $(2r + t)^7$

5. $(2p + 3m)^6$

6. $\left(a - \dfrac{b}{2}\right)^4$

7. **COIN TOSS** Ray tosses a coin 15 times. How many different sequences of tosses could result in 4 heads and 11 tails?

8. **QUIZZES** There are 9 true/false questions on a quiz. If twice as many of the statements are true as false, how many different sequences of true/false answers are possible?

11-6 Study Guide and Intervention *(continued)*

The Binomial Theorem

The Binomial Theorem

Binomial Theorem	If n is a natural number, then $(a + b)^n =$ $_nC_0 a^n b^0 + {}_nC_1 a^{n-1}b^1 + {}_nC_2 a^{n-2}b^2 + \cdots + {}_nC_n a^0 b^n = \displaystyle\sum_{k=0}^{n} \frac{n!}{k!(n-k)!} a^{n-k}b^k.$

Example Expand $(a - 3b)^4$.

$$(a - 3b)^4 = \sum_{k=0}^{4} \frac{4!}{(4-k)!k!} a^{4-k}(-3b)^k$$

$$= \frac{4!}{4!0!}a^4 + \frac{4!}{3!1!}a^3(-3b)^1 + \frac{4!}{2!2!}a^2(-3b)^2 + \frac{4!}{1!3!}a(-3b)^3 + \frac{4!}{0!4!}(-3b)^4$$

$$= a^4 - 12a^3b + 54a^2b^2 - 108ab^3 + 81b^4$$

Exercises

Expand each binomial.

1. $(a - 3)^6$

2. $(r + 2t)^7$

3. $(4x + y)^4$

4. $\left(2 - \dfrac{m}{2}\right)^5$

Find the indicated term of each expansion.

5. third term of $(3x - y)^5$

6. fifth term of $(a + 1)^7$

7. fourth term of $(j + 2k)^8$

8. sixth term of $(10 - 3t)^7$

9. second term of $\left(m + \dfrac{2}{3}\right)^9$

10. seventh term of $(5x - 2)^{11}$

11-7 Study Guide and Intervention

Proof by Mathematical Induction

Mathematical Induction Mathematical induction is a method of proof used to prove statements about natural numbers.

Mathematical Induction Proof	Step 1	Show that the statement is true for some integer n.
	Step 2	Assume that the statement is true for some positive integer k where $k \geq n$. This assumption is called the **induction hypothesis**.
	Step 3	Show that the statement is true for the next integer $k + 1$.

Example   Prove that $5 + 11 + 17 + \ldots + (6n - 1) = 3n^2 + 2n$.

Step 1 When $n = 1$, the left side of the given equation is $6(1) - 1 = 5$. The right side is $3(1)^2 + 2(1) = 5$. Thus the equation is true for $n = 1$.

Step 2 Assume that $5 + 11 + 17 + \ldots + (6k - 1) = 3k^2 + 2k$ for some positive integer k.

Step 3 Show that the equation is true for $n = k + 1$. First, add $[6(k + 1) - 1]$ to each side.

$$5 + 11 + 17 + \ldots + (6k - 1) + [6(k + 1) - 1] = 3k^2 + 2k + [6(k + 1) - 1]$$

$$\begin{aligned} &= 3k^2 + 2k + 6k + 5 && \text{Add.} \\ &= 3k^2 + 6k + 3 + 2k + 2 && \text{Rewrite.} \\ &= 3(k^2 + 2k + 1) + 2(k + 1) && \text{Factor.} \\ &= 3(k + 1)^2 + 2(k + 1) && \text{Factor.} \end{aligned}$$

The last expression above is the right side of the equation to be proved, where n has been replaced by $k + 1$. Thus the equation is true for $n = k + 1$.

This proves that $5 + 11 + 17 + \ldots + (6n - 1) = 3n^2 + 2n$ for all positive integers n.

Exercises

Prove that each statement is true for all natural numbers.

1. $3 + 7 + 11 + \ldots + (4n - 1) = 2n^2 + n$.

2. $500 + 100 + 20 + \ldots + 4 \cdot 5^{4-n} = 625\left(1 - \dfrac{1}{5^n}\right)$.

11-7 Study Guide and Intervention *(continued)*

Proof by Mathematical Induction

Counterexamples To show that a formula or other generalization is *not* true, find a counterexample. Often this is done by substituting values for a variable.

Example 1 Find a counterexample to disprove the formula $2n^2 + 2n + 3 = 2^{n+2} - 1$.

Check the first few positive integers.

n	Left Side of Formula	Right Side of Formula	
1	$2(1)^2 + 2(1) + 3 = 2 + 2 + 3$ or 7	$2^{1+2} - 1 = 2^3 - 1$ or 7	true
2	$2(2)^2 + 2(2) + 3 = 8 + 4 + 3$ or 15	$2^{2+2} - 1 = 2^4 - 1$ or 15	true
3	$2(3)^2 + 2(3) + 3 = 18 + 6 + 3$ or 27	$2^{3+2} - 1 = 2^5 - 1$ or 31	false

Example 2 Find a counterexample to disprove the statement $x^2 + 4$ is either prime or divisible by 4.

n	$x^2 + 4$	True?	n	$x^2 + 4$	True?
1	1 + 4 or 5	Prime	6	36 + 4 or 40	Div. by 4
2	4 + 4 or 8	Div. by 4	7	49 + 4 or 53	Prime
3	9 + 4 or 13	Prime	8	64 + 4 or 68	Div. by 4
4	16 + 4 or 20	Div. by 4	9	81 + 4 or 85	Neither
5	25 + 4 or 29	Prime			

The value $n = 9$ provides a counterexample.

Exercises

Find a counterexample to disprove each statement.

1. $1 + 5 + 9 + \ldots + (4n - 3) = 4n - 3$

2. $100 + 110 + 120 + \ldots + (10n + 90) = 5n^2 + 95$

3. $900 + 300 + 100 + \ldots + 100(3^{3-n}) = 900 \cdot \dfrac{2n}{n+1}$

4. $n^2 + n + 1$ is prime.

5. $2n + 1$ is a prime number.

6. $7n - 5$ is a prime number.

7. $\dfrac{1}{2} + 1 + \dfrac{3}{2} + \ldots + \dfrac{n}{2} = n - \dfrac{1}{2}$

8. $5n^2 + 1$ is divisible by 3.

9. $n^2 - 3n + 1$ is prime for $n > 2$.

10. $4n^2 - 1$ is divisible by either 3 or 5.

12-1 Study Guide and Intervention

Experiments, Surveys, and Observational Studies

Surveys, Studies, and Experiments

Term	Definition	Example
Survey	a means of obtaining information from a population or a sample of the population	taking a poll to learn who people will vote for in an upcoming election
Observational Study	an examination in which individuals are observed and no attempt is made to influence the results	observing a group of 100 people, 50 of whom have been taking a treatment; collecting, analyzing, and interpreting the data
Experiment	an operation in which something is intentionally done to people, animals, or objects, and then the response is observed	studying the differences between two groups of people, one of which receives a treatment and the other a placebo

Example 1 State whether the following situation represents an *experiment* or an *observational study*. If it is an experiment, then identify the control group and treatment group. Then determine whether there is bias.

Find twenty males and randomly separate them into two groups. One group will receive a new trial medication and the other receives a placebo.

This is an experiment. The group receiving the medication is the treatment group, while the group receiving the placebo is the control group. This is unbiased.

Example 2 Determine whether the following situation calls for a *survey*, an *observational study*, or an *experiment*. Explain the process.

You want to know how students and parents feel about school uniforms.

This calls for a survey. It is best to ask a random sample of students and a random sample of parents to give their opinions.

Exercises

State whether each situation represents an *experiment* or an *observational study*. If it is an experiment, then identify the control group and the treatment group. Then determine whether there is bias.

1. Find 300 students and randomly split them into two groups. One group practices basketball three times per week and the other group does not practice basketball at all. After three months, you interview the students to find out how they feel about school.

2. Find 100 students, half of whom participated on the school math team, and compare their grade point average.

12-1 Study Guide and Intervention *(continued)*

Experiments, Surveys, and Observational Studies

Distinguish Between Correlation and Causation

Term	Definition	Example
correlation	When one event happens, the other is more likely.	When the pond is frozen, it is more likely to snow.
causation	One event is the direct cause of another event.	Turning on the light makes a room brighter.

Example **Determine whether the following statement shows *correlation* or *causation*. Explain your reasoning.**

Children who live in very large houses usually get larger allowances than children who live in small houses.

Correlation; there is no reason to assume that the size of their house causes children to receive more allowance. Children living in both a large house and getting a large allowance could be the result of a third factor—the amount of money the parents have.

Exercises

Determine whether the following statements show *correlation* or *causation*. Explain.

1. If I jog in the rain, I will get sick.

2. Studies have shown that eating more fish will improve your math grade.

3. If you lose a library book, you will have to pay a fine.

4. Reading a diet book will make you lose weight.

5. If I miss a day of school, I will not earn the perfect attendance award.

6. Owning an expensive car will make me earn lots of money.

12-2 Study Guide and Intervention

Statistical Analysis

Measures of Central Tendency

Term	Definition	Best Used When
Mean	sum of the data divided by number of items in the data	the data set has no outliers
Median	the middle number or mean of two middle numbers of the ordered data	the data set has outliers but no big gaps in the middle of the data
Mode	the number or numbers that occur most often	the data has many repeated numbers
Margin of Sampling Error	$\pm\dfrac{1}{\sqrt{n}}$, for a random sample of n items	estimating how well a sample represents the whole population

Example 1 **Which measure of central tendency best represents the data and why?**

{2.1, 21.5, 22.3, 22.8, 23.1, 159.4}

There are outliers, but no large gaps in the middle, the median best represents this data.

Example 2 **What is the margin of sampling error and the likely interval that contains the percentage of the population?**

Of 400 people surveyed in a national poll, 51% say they will vote for candidate González.

Since 400 people are surveyed, the margin of sampling error is $\pm\dfrac{1}{\sqrt{400}}$ or $\pm 5\%$. The percentage of people who are likely to vote for candidate González is the percentage found in the survey plus or minus 5%, so the likely interval is from 46% to 56%.

Exercises

Which measure of central tendency best represents the data, and why?

1. {45, 16, 30, 45, 29, 45}

2. {100, 92, 105, 496, 77, 121, 65, 99, 72}

3. {2.5, 99.5, 110.5, 76, 88.5, 105, 73, 113, 92, 72.5}

4. {60, 50, 55, 62, 44, 65, 51}

5. **BOOKS** A survey of 28 random people found that 40% read at least three books each month. What is the margin of sampling error? What is the likely interval that contains the percentage of the population that reads at least three books each month?

12-2 Study Guide and Intervention *(continued)*

Statistical Analysis

Measures of Variation

Standard Deviation Formulas		
	Variable	Formula
For Samples	s	$\sqrt{\dfrac{\sum\limits_{k=1}^{n}(x_k - \bar{x})^2}{n-1}}$
For Populations	σ	$\sqrt{\dfrac{\sum\limits_{k=1}^{n}(x_k - \mu)^2}{n}}$

Example **For the following data, determine whether it is a sample or a population. Then find the standard deviation of the data. Round to the nearest hundredth.**

The test scores of the twelve students in a college mathematics course are displayed below.

Test Scores of Twelve Students Enrolled in a College Mathematics Course					
61	99	75	83	92	69
77	94	73	65	98	91

Because the scores of all 12 students enrolled are given, this is a population. Find the mean.

$$\mu = \frac{\sum\limits_{n=1}^{12} Xn}{12} = \frac{977}{12} \approx 81.42$$

Next, take the sum of the squares of the differences between each score and the mean.

$$\Sigma\left[(61 - 81.42)^2 + (99 - 81.42)^2 + (75 - 81.42)^2 + \dots + (91 - 81.42)^2\right] \approx 1920.92$$

Putting this into the standard deviation formula, $\sqrt{\dfrac{1920.92}{12}} \approx 12.65$

Exercises

1. Determine whether each is a sample or a population. Then find the standard deviation of the data. Round to the nearest hundredth.

a.

The Test Scores of Some of the Females in a College History Course					
88	91	82	95	76	88
75	94	92	85	82	90

b.

The Age of All Students in the Chess Club					
14	17	15	14	15	16

12-3 Study Guide and Intervention
Conditional Probability

Conditional Probability The probability of an event, given that another event has already occurred, is called **conditional probability**. The conditional probability that event B occurs, given that event A has already occurred, can be represented by $P(B \mid A)$.

Example   SPINNER Naomi is playing a game with the spinner shown. What is the probability that the spinner lands on 7, given that Naomi has spun a number greater than 5?

There are 8 possible results of spinning the spinner shown.

Let event A be that she spun a number greater than 5.

Let event B be that she spun a 7.

$P(A)$	$= \dfrac{3}{8}$	Three of the eight outcomes are greater than 5.
$P(A \text{ and } B)$	$= \dfrac{1}{8}$	One out of eight outcomes is both greater than 5 and equal to 7.
$P(B \mid A)$	$= \dfrac{P(A \text{ and } B)}{P(B)}$	Conditional Probability Formula
$P(B \mid A)$	$= \dfrac{1}{8} \div \dfrac{3}{8}$	Substitute values for $P(A)$ and $P(A \text{ and } B)$
$P(B \mid A)$	$= \dfrac{1}{3}$	

The probability of spinning a 7, given that the spin is greater than 5, is $\dfrac{1}{3}$.

Exercises

Cards are drawn from a standard deck of 52 cards. Find each probability.

1. The card is a heart, given that an ace is drawn.

2. The card is the six of clubs, given that a club is drawn.

3. The card is a spade, given that a black card is drawn. (Hint: The black cards are the suits of clubs and spades.)

A six-sided die is rolled. Find each probability.

4. A 4 is rolled, given that the roll is even.

5. A 2 is rolled, given that the roll is less than 6.

6. An even number is rolled, given that the roll is 3 or 4.

12-3 Study Guide and Intervention (continued)

Conditional Probability

Contingency Tables A **contingency table** is a table that records data in which different possible situations result in different possible outcomes. These tables can be used to determine conditional probabilities.

Example **LACROSSE** Find the probability that a student plays lacrosse, given that the student is a junior.

Class	Freshman	Sophomore	Junior	Senior
Plays Lacrosse	17	20	34	17
Does Not Play	586	540	510	459

There are a total of $14 + 586 + 20 + 540 + 34 + 510 + 17 + 459 = 2180$ students.

$$P(L \mid J) = \frac{P(L \text{ and } J)}{P(J)} \qquad \text{Conditional Probability Formula}$$

$$= \frac{34}{2180} \div \frac{544}{2180} \qquad P(L \text{ and } J) = \frac{34}{2180} \text{ and } P(J) = \frac{34 + 510}{2180}$$

$$= \frac{34}{544} \text{ or } \frac{1}{16} \qquad \text{Simplify.}$$

The probability that a student plays lacrosse, given that the student is a junior, is $\frac{1}{16}$.

Exercises

1. **WEDDINGS** The table shows attendance at a wedding. Find the probability that a person can attend the wedding, given that the person is in the bride's family.

Family	Bride's Family	Groom's Family
Can Attend	104	112
Cannot Attend	32	14

2. **BASEBALL** The table shows the number of students who play baseball. Find the probability that a student plays baseball, given that the student is a senior.

Class	Plays Baseball	Does Not Play Baseball
Junior	22	352
Senior	34	306

3. **SHOPPING** Four businesses in town responded to a questionnaire asking how many people paid for their purchases last month using cash, credit cards, and debit cards. Find each probability.

Class	Jacob's Gas	Gigantomart	T.J.'s	Pet Town
Cash	304	140	102	49
Credit Card	456	223	63	70
Debit Card	380	166	219	28

a. A shopper uses a credit card, given that the shopper is shopping at Jacob's Gas.

b. A shopper uses a debit card, given that the shopper is shopping at Pet Town.

c. A shopper is shopping at T.J.'s, given that the shopper is paying cash.

12-4 Study Guide and Intervention
Probability and Probability Distributions

Probability In probability, a desired outcome is called a **success**; any other outcome is called a **failure**.

Probability of Success and Failure	If an event can succeed in s ways and fail in f ways, then the probabilities of success, $P(S)$, and of failure, $P(F)$, are as follows.$$P(S) = \frac{s}{s + f} \text{ and } P(F) = \frac{f}{s + f}$$

Example 1 When 3 coins are tossed, what is the probability that at least 2 are heads?

You can use a tree diagram to find the sample space.

First Coin	Second Coin	Third Coin	Possible Outcomes
	H	H	HHH
		T	HHT
H	T	H	HTH
		T	HTT
	H	H	THH
		T	THT
T	T	H	TTH
		T	TTT

Of the 8 possible outcomes, 4 have at least 2 heads. So the probability of tossing at least 2 heads is $\frac{4}{8}$ or $\frac{1}{2}$.

Example 2 What is the probability of picking 4 fiction books and 2 biographies from a best-seller list that consists of 12 fiction books and 6 biographies?

The number of successes is $_{12}C_4 \cdot \, _6C_2$.
The total number of selections, $s + f$, of 6 books is $C(18, 6)$.

$$P(4 \text{ fiction}, 2 \text{ biography}) = \frac{_{12}C_4 \cdot \, _6C_2}{_{18}C_6} \text{ or about } 0.40$$

The probability of selecting 4 fiction books and 2 biographies is about 40%.

Exercises

1. **PET SHOW** A family has 3 dogs and 4 cats. Find the probability of each of the following if they select 2 pets at random to bring to a local pet show.

   a. $P(2 \text{ cats})$ **b.** $P(2 \text{ dogs})$ **c.** $P(1 \text{ cat, 1 dog})$

2. **MUSIC** Eduardo's MP3 player has 10 blues songs and 5 rock songs (and no other music). Find the probability of each of the following if he plays six songs at random and songs may not repeat . Round to the nearest tenth of a percent.

   a. $P(6 \text{ blues songs})$ **b.** $P(4 \text{ blues songs, 2 rock songs})$ **c.** $P(2 \text{ blues songs, 4 rock songs})$

3. **CANDY** One bag of candy contains 15 red candies, 10 yellow candies, and 6 green candies. Find the probability of each selection.

   a. picking a red candy   **b.** not picking a yellow candy

   c. picking a green candy   **d.** not picking a red candy

12-4 Study Guide and Intervention *(continued)*

Probability and Probability Distributions

Probability Distributions A **random variable** is a variable whose value is the numerical outcome of a random event. A **probability distribution** for a particular random variable is a function that maps the sample space to the probabilities of the outcomes in the sample space.

Example Suppose two dice are rolled. The table and the relative-frequency graph show the distribution of the absolute value of the difference of the numbers rolled. Use the graph to determine which outcome is the most likely. What is its probability?

Difference	0	1	2	3	4	5
probability	$\frac{1}{6}$	$\frac{5}{18}$	$\frac{2}{9}$	$\frac{1}{6}$	$\frac{1}{9}$	$\frac{1}{18}$

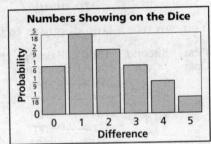

Numbers Showing on the Dice

The greatest probability in the graph is $\frac{5}{18}$.

The most likely outcome is a difference of 1 and its probability is $\frac{5}{18}$.

Exercises

1. **PROBABILITY** Four coins are tossed.

   a. Complete the table below to show the probability distribution of the number of heads.

Number of Heads	0	1	2	3	4
Probability					

   b. Create a relative-frequency graph.

   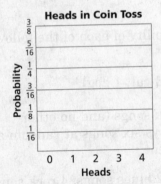

   Heads in Coin Toss

   c. Find P(four heads).

12-5 Study Guide and Intervention

The Normal Distribution

Normal and Skewed Distributions A continuous probability distribution is represented by a curve.

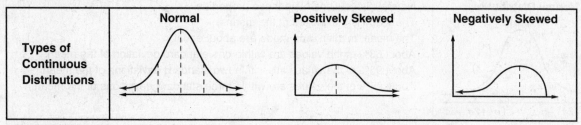

Types of Continuous Distributions	Normal	Positively Skewed	Negatively Skewed

Example Determine whether the data below appear to be *positively skewed*, *negatively skewed*, or *normally distributed*.

{100, 120, 110, 100, 110, 80, 100, 90, 100, 120, 100, 90, 110, 100, 90, 80, 100, 90}

Make a frequency table for the data.

Value	80	90	100	110	120
Frequency	2	4	7	3	2

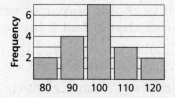

Then use the data to make a graph.

Since the graph is roughly symmetric, the data appear to be normally distributed.

Exercises

Determine whether the data appear to be *positively skewed*, *negatively skewed*, or *normally distributed*. Make a graph of the data.

1. {27, 24, 29, 25, 27, 22, 24, 25, 29, 24, 25, 22, 27, 24, 22, 25, 24, 22}

2.

Shoe Size	4	5	6	7	8	9	10
No. of Students	1	2	4	8	5	1	2

3.

Housing Price	No. of Houses Sold
less than $100,000	0
$100,00–$120,000	1
$121,00–$140,000	3
$141,00–$160,000	7
$161,00–$180,000	8
$181,00–$200,000	6
over $200,000	12

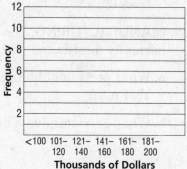

12-5   Study Guide and Intervention   *(continued)*

The Normal Distribution

The Empirical Rule

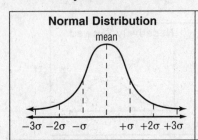

Normal Distribution

	Normal distributions have these properties.
	The graph is maximized at the mean.
	The mean, median, and mode are about equal.
	About 68% of the values are within one standard deviation of the mean.
	About 95% of the values are within two standard deviations of the mean.
	About 99% of the values are within three standard deviations of the mean.

Example   The heights of players in a basketball league are normally distributed with a mean of 6 feet 1 inch and a standard deviation of 2 inches.

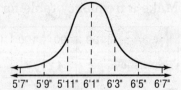

a. What is the probability that a player selected at random will be shorter than 5 feet 9 inches?

Draw a normal curve. Label the mean and the mean plus or minus multiples of the standard deviation.

The value of 5 feet 9 inches is 2 standard deviations below the mean, so approximately 2.5% of the players will be shorter than 5 feet 9 inches.

b. If there are 240 players in the league, about how many players are taller than 6 feet 3 inches?

The value of 6 feet 3 inches is one standard deviation above the mean. Approximately 16% of the players will be taller than this height.

$240 \times 0.16 \approx 38$

About 38 of the players are taller than 6 feet 3 inches.

Exercises

1. **EGG PRODUCTION** The number of eggs laid per year by a particular breed of chicken is normally distributed with a mean of 225 and a standard deviation of 10 eggs.

   a. About what percent of the chickens will lay between 215 and 235 eggs per year?

   b. In a flock of 400 chickens, about how many would you expect to lay more than 245 eggs per year?

2. **MANUFACTURING** The diameter of bolts produced by a manufacturing plant is normally distributed with a mean of 18 mm and a standard deviation of 0.2 mm.

   a. What percent of bolts coming off of the assembly line have a diameter greater than 18.4 mm?

   b. What percent have a diameter between 17.8 and 18.2 mm?

12-6 Study Guide and Intervention

Hypothesis Testing

Confidence Interval

Term	Definition
Confidence Interval	the estimated range within which a number will fall with a stated degree of certainty
95% Confidence Interval Formula	$CI = \bar{x} \pm 2 \cdot \dfrac{s}{\sqrt{n}}$

Example A survey asked 100 random people how many minutes they exercised each day. The mean of their answers was 25.3 minutes with a standard deviation of 9.4 minutes. Determine a 95% confidence interval. Round to the nearest tenth.

$CI = \bar{x} \pm 2 \cdot \dfrac{s}{\sqrt{n}}$ Confidence Interval Formula

$= 25.3 \pm 2 \cdot \dfrac{9.4}{\sqrt{100}}$ $\bar{x} = 25.3,\ s = 9.4,\ n = 100$

$= 25.3 \pm 1.88$

The 95% confidence interval to the nearest tenth is $23.4 \le \mu \le 27.2$.

Exercises

Find a 95% confidence interval for each of the following. Round to the nearest tenth if necessary.

1. $\bar{x} = 10$, $s = 6$, and $n = 100$

2. $\bar{x} = 100$, $s = 5.4$, and $n = 5$

3. $\bar{x} = 90$, $s = 1.8$, and $n = 170$

4. $\bar{x} = 82$, $s = 4.5$, and $n = 8000$

5. $\bar{x} = 1088$, $s = 7.8$, and $n = 200$

6. $\bar{x} = 70$, $s = 10$, and $n = 50$

7. $\bar{x} = 120$, $s = 8$, and $n = 1000$

8. $\bar{x} = 147$, $s = 39$, and $n = 100$

9. $\bar{x} = 70.5$, $s = 5.5$, and $n = 150$

10. $\bar{x} = 788.2$, $s = 52$, and $n = 8$

12-6 Study Guide and Intervention *(continued)*

Hypothesis Testing

Hypothesis Testing

Steps in Testing a Null Hypothesis	
Step 1	State the null hypothesis H_0 and alternative hypothesis H_1.
Step 2	Design the experiment.
Step 3	Conduct the experiment and collect the data.
Step 4	Find the confidence interval.
Step 5	Make the correct statistical inference. Accept the null hypothesis if the population parameter falls within the confidence interval.

Example A team of students claimed the average student at their school studied at least 27.5 hours per week. They designed an experiment using the 5 steps for testing a null hypothesis.

Step 1 The null hypothesis H_0: $\mu \leq 27.5$ hours per week.
The alternative hypothesis H_1: $\mu > 27.5$ hours per week.

Step 2 They decided to survey students and wrote a questionnaire.

Step 3 They surveyed 10 students. They found $\bar{x} = 29.2$ and $s = 5.3$.

Step 4 They calculated the confidence interval.

$$CI = \bar{x} \pm 2 \cdot \frac{s}{\sqrt{n}} \qquad \text{Confidence Interval Formula}$$

$$= 29.2 \pm 2 \cdot \frac{5.3}{\sqrt{10}} \qquad \bar{x} = 25.3, s = 9.4, n = 100$$

$$= 29.2 \pm 3.35 \qquad \text{Use a calculator.}$$

Step 5 The null hypothesis H_0 overlaps the confidence interval, so the students should accept the null hypothesis. They have not proven that the average student at their school studies at least 27.5 hours per week.

Exercises

Test each null hypothesis. Write *accept* or *reject*.

1. $H_0 = 20$, $H_1 < 20$, $n = 50$, $\bar{x} = 12$, and $s = 2$

2. $H_0 = 21$, $H_1 < 21$, $n = 100$, $\bar{x} = 20$, and $s = 5$

3. $H_0 = 80$, $H_1 > 80$, $n = 50$, $\bar{x} = 80$, and $s = 21.3$

4. $H_0 = 64.5$, $H_1 > 64.5$, $n = 150$, $\bar{x} = 68$, and $s = 3.5$

5. $H_0 = 87.6$, $H_1 > 87.6$, $n = 1200$, $\bar{x} = 88$, and $s = 2$

6. $H_0 = 10.4$, $H_1 < 10.4$, $n = 200$, $\bar{x} = 10$, and $s = 4$

7. $H_0 = 99.44$, $H_1 > 99.44$, $n = 10$, $\bar{x} = 100$, and $s = 2$

8. $H_0 = 16.2$, $H_1 < 16.2$, $n = 150$, $\bar{x} = 15.8$, and $s = 3.5$

9. $H_0 = 49.2$, $H_1 < 49.2$, $n = 100$, $\bar{x} = 49$, and $s = 1$

10. $H_0 = 298$, $H_1 > 298$, $n = 225$, $\bar{x} = 300$, and $s = 15$

12-7 Study Guide and Intervention

Binomial Distributions

Binomial Experiments

Binomial Experiments	A binomial experiment is possible if and only if all of these conditions occur. • There are exactly two outcomes for each trial. • There is a fixed number of trials. • The trials are independent. • The probabilities for each trial are the same.

Example Suppose a coin is weighted so that the probability of getting heads in any one toss is 90%. What is the probability of getting exactly 7 heads in 8 tosses?

The probability of getting heads is $\frac{9}{10}$ and the probability of getting tails is $\frac{1}{10}$. There are $C(8, 7)$ ways to choose the 7 heads.

$$P(7 \text{ heads}) = C(8, 7)\left(\frac{9}{10}\right)^7\left(\frac{1}{10}\right)^1$$
$$= 8 \cdot \frac{9^7}{10^8}$$
$$\approx 0.38$$

The probability of getting 7 heads in 8 tosses is about 38%.

Exercises

1. **BASKETBALL** For any one foul shot, Derek has a probability of 0.72 of getting the shot in the basket. As part of a practice drill, he shoots 8 shots from the foul line.

   a. What is the probability that he gets in exactly 6 foul shots?

   b. What is the probability that he gets in at least 6 foul shots?

2. **SCHOOL** A teacher is trying to decide whether to have 4 or 5 choices per question on her multiple choice test. She wants to prevent students who just guess from scoring well on the test.

   a. On a 5-question multiple-choice test with 4 choices per question, what is the probability that a student can score at least 60% by guessing?

   b. What is the probability that a student can score at least 60% by guessing on a test of the same length with 5 choices per question?

3. **DICE** Julie rolls two dice and adds the two numbers.

   a. What is the probability that the sum will be divisible by 3?

   b. If she rolls the dice 5 times what is the chance that she will get exactly 3 sums that are divisible by 3?

4. **SKATING** During practice a skater falls 15% of the time when practicing a triple axel. During one practice session he attempts 20 triple axels.

   a. What is the probability that he will fall only once?

   b. What is the probability that he will fall 4 times?

12-7 Study Guide and Intervention *(continued)*

Binomial Distributions

Binomial Distribution For situations with only 2 possible outcomes, you can use the Binomial Theorem to find probabilities. The coefficients of terms in a binomial expansion can be found by using combinations.

Binomial Theorem for Probabilities
The probability of x successes in n independent trials is $P(x) = C(n, x) \, s^x f^{n-x}$, where s is the probability of success of an individual trial and f is the probability of failure on that same individual trial. $(s + f = 1.)$

Example **What is the probability that 3 coins show heads and 3 show tails when 6 coins are tossed?**

There are 2 possible outcomes that are equally likely: heads (H) and tails (T). The tosses of 6 coins are independent events. When $(H + T)^6$ is expanded, the term containing H^3T^3, which represents 3 heads and 3 tails, is used to get the desired probability. By the Binomial Theorem the coefficient of H^3T^3 is $C(6, 3)$.

$$P(3 \text{ heads}, 3 \text{ tails}) = \frac{6!}{3!3!}\left(\frac{1}{2}\right)^3\left(\frac{1}{2}\right)^3 \qquad P(H) = \frac{1}{2} \text{ and } P(T) = \frac{1}{2}$$

$$= \frac{20}{64}$$

$$= \frac{5}{16}$$

The probability of getting 3 heads and 3 tails is $\frac{5}{16}$ or 0.3125.

Exercises

1. **COINS** Find each probability if a coin is tossed 8 times.

   a. P(exactly 5 heads)

   b. P(exactly 2 heads)

   c. P(even number of heads)

   d. P(at least 6 heads)

2. **TRUE-FALSE** Mike guesses on all 10 questions of a true-false test. If the answers true and false are evenly distributed, find each probability.

   a. Mike gets exactly 8 correct answers.

   b. Mike gets at most 3 correct answers.

3. A die is tossed 4 times. What is the probability of tossing exactly two sixes?

13-1 Study Guide and Intervention

Trigonometric Functions in Right Triangles

Trigonometric Functions for Acute Angles Trigonometry is the study of relationships among the angles and sides of a right triangle. A **trigonometric function** has a rule given by a **trigonometric ratio**, which is a ratio that compares the side lengths of a right triangle.

Trigonometric Functions in Right Triangles	If θ is the measure of an acute angle of a right triangle, *opp* is the measure of the leg opposite θ, *adj* is the measure of the leg adjacent to θ, and *hyp* is the measure of the hypotenuse, then the following are true.
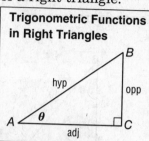	$\sin \theta = \dfrac{opp}{hyp}$ $\qquad \cos \theta = \dfrac{adj}{hyp} \qquad \tan \theta = \dfrac{opp}{adj}$ $\csc \theta = \dfrac{hyp}{opp} \qquad \sec \theta = \dfrac{hyp}{adj} \qquad \cot \theta = \dfrac{adj}{opp}$

Example In a right triangle, $\angle B$ is acute and $\cos B = \dfrac{3}{7}$. Find the value of $\tan B$.

Step 1 Draw a right triangle and label one acute angle B. Label the adjacent side 3 and the hypotenuse 7.

Step 2 Use the Pythagorean Theorem to find b.

$a^2 + b^2 = c^2$ Pythagorean Theorem

$3^2 + b^2 = 7^2$ $a = 3$ and $c = 7$

$9 + b^2 = 49$ Simplify.

$b^2 = 40$ Subtract 9 from each side.

$b = \sqrt{40} = 2\sqrt{10}$ Take the positive square root of each side.

Step 3 Find $\tan B$.

$\tan B = \dfrac{opp}{adj}$ Tangent function

$\tan B = \dfrac{2\sqrt{10}}{3}$ Replace *opp* with $2\sqrt{10}$ and *adj* with 3.

Exercises

Find the values of the six trigonometric functions for angle θ.

1.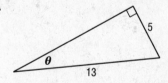

2.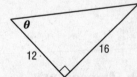

3.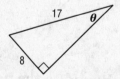

In a right triangle, $\angle A$ and $\angle B$ are acute.

4. If $\tan A = \dfrac{7}{12}$, what is $\cos A$?

5. If $\cos A = \dfrac{1}{2}$, what is $\tan A$?

6. If $\sin B = \dfrac{3}{8}$, what is $\tan B$?

13-1 Study Guide and Intervention (continued)

Trigonometric Functions in Right Triangles

Use Trigonometric Functions You can use trigonometric functions to find missing side lengths and missing angle measures of right triangles. You can find the measure of the missing angle by using the inverse of sine, cosine, or tangent.

Example Find the measure of $\angle C$. Round to the nearest tenth if necessary.

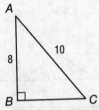

You know the measure of the side opposite $\angle C$ and the measure of the hypotenuse. Use the sine function.

$\sin C = \dfrac{\text{opp}}{\text{hyp}}$ Sine function

$\sin C = \dfrac{8}{10}$ Replace *opp* with 8 and *hyp* with 10.

$\sin^{-1} \dfrac{8}{10} = m\angle C$ Inverse sine

$53.1° \approx m\angle C$ Use a calculator.

Exercises

Use a trigonometric function to find each value of x. Round to the nearest tenth if necessary.

1.
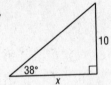

2.
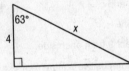

3.
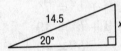

4.

5.

6.
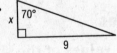

Find x. Round to the nearest tenth if necessary.

7.

8.
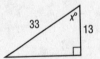

9.
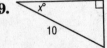

13-2 Study Guide and Intervention

Angles and Angle Measure

Angles in Standard Position An angle is determined by two rays. The degree measure of an angle in standard position is described by the amount and direction of rotation from the **initial side**, which lies along the positive x-axis, to the **terminal side**. A counterclockwise rotation is associated with positive angle measure and a clockwise rotation is associated with negative angle measure. Two or more angles in standard position with the same terminal side are called **coterminal angles**.

Example 1 Draw an angle with measure 290° in standard position.

The negative y-axis represents a positive rotation of 270°. To generate an angle of 290°, rotate the terminal side 20° more in the counterclockwise direction

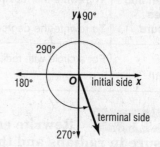

Example 2 Find an angle with a positive measure and an angle with a negative measure that are coterminal with each angle.

a. 250°

A positive angle is 250° + 360° or 610°.   Add 360°.

A negative angle is 250° − 360° or −110°.   Subtract 360°.

b. −140°

A positive angle is −140° + 360° or 220°.   Add 360°.

A negative angle is −140° − 360° or −500°.   Subtract 360°.

Exercises

Draw an angle with the given measure in standard position.

1. 160°   **2.** 280°   **3.** 400°

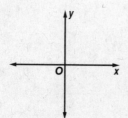   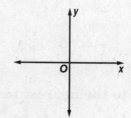   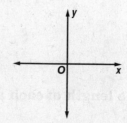

Find an angle with a positive measure and an angle with a negative measure that are coterminal with each angle.

4. 65°   **5.** −75°   **6.** 230°   **7.** 420°

13-2 Study Guide and Intervention (continued)

Angles and Angle Measure

Convert Between Degrees and Radians Angles can be measured in **degrees** and **radians**, which are units based on arc length. One radian is the measure of an angle θ in standard position with a terminal side that intercepts an arc with the same length as the radius of the circle. Degree measure and radian measure are related by the equations 2π radians $= 360°$ and π radians $= 180°$.

Radian and Degree Measure	To rewrite the radian measure of an angle in degrees, multiply the number of radians by $\frac{180°}{\pi \text{ radians}}$.
	To rewrite the degree measure of an angle in radians, multiply the number of degrees by $\frac{\pi \text{ radians}}{180°}$.
Arc Length	For a circle with radius r and central angle θ (in radians), the arc length s equals the product of r and θ. $$s = r\theta$$

Example 1 Rewrite each degree measure in radians and the radian measure in degrees.

a. **45°**

$$45° = 45°\left(\frac{\pi \text{ radians}}{180°}\right) = \frac{\pi}{4} \text{ radians}$$

b. $\frac{5\pi}{3}$ **radians**

$$\frac{5\pi}{3} \text{ radians} = \frac{5\pi}{3}\left(\frac{180°}{\pi}\right) = 300°$$

Example 2 A circle has a radius of 5 cm and central angle of 135°, what is the length of the circle's arc?

Find the central angle in radians.

$$135° = 135°\left(\frac{\pi \text{ radians}}{180°}\right) = \frac{3\pi}{4} \text{ radians}$$

Use the radius and central angle to find the arc length.

$s = r\theta$	Write the formula for arc length.
$= 5 \cdot \dfrac{3\pi}{4}$	Replace r with 5 and θ with $\frac{3\pi}{4}$.
≈ 11.78	Use a calculator to simplify.

Exercises

Rewrite each degree measure in radians and each radian measure in degrees.

1. 140°

2. −260°

3. $-\dfrac{3\pi}{5}$

4. −75°

5. $\dfrac{7\pi}{6}$

6. 380°

Find the length of each arc. Round to the nearest tenth.

7.

8.

9.

13-3　Study Guide and Intervention

Trigonometric Functions of General Angles

Trigonometric Functions for General Angles

Trigonometric Functions, θ in Standard Position	Let θ be an angle in standard position and let $P(x, y)$ be a point on the terminal side of θ. By the Pythagorean Theorem, the distance r from the origin is given by $r = \sqrt{x^2 + y^2}$. The trigonometric functions of an angle in standard position may be defined as follows.
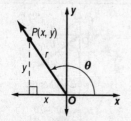	$\sin \theta = \dfrac{y}{r}$　　$\cos \theta = \dfrac{x}{r}$　　$\tan \theta = \dfrac{y}{x}, x \neq 0$
	$\csc \theta = \dfrac{r}{y}, y \neq 0$　$\sec \theta = \dfrac{r}{x}, x \neq 0$　$\cot \theta = \dfrac{x}{y}, y \neq 0$

Example　　Find the exact values of the six trigonometric functions of θ if the terminal side of θ in standard position contains the point $(-5, 5\sqrt{2})$.

You know that $x = -5$ and $y = 5$. You need to find r.

$r = \sqrt{x^2 + y^2}$　　Pythagorean Theorem

$= \sqrt{(-5)^2 + (5\sqrt{2})^2}$　　Replace x with -5 and y with $5\sqrt{2}$.

$= \sqrt{75}$ or $5\sqrt{3}$

Now use $x = -5$, $y = 5\sqrt{2}$, and $r = 5\sqrt{3}$ to write the six trigonometric ratios.

$\sin \theta = \dfrac{y}{r} = \dfrac{5\sqrt{2}}{5\sqrt{3}} = \dfrac{\sqrt{6}}{3}$　　$\cos \theta = \dfrac{x}{r} = \dfrac{-5}{5\sqrt{3}} = -\dfrac{\sqrt{3}}{3}$　　$\tan \theta = \dfrac{y}{x} = \dfrac{5\sqrt{2}}{-5} = -\sqrt{2}$

$\csc \theta = \dfrac{r}{y} = \dfrac{5\sqrt{3}}{5\sqrt{2}} = \dfrac{\sqrt{6}}{2}$　　$\sec \theta = \dfrac{r}{x} = \dfrac{5\sqrt{3}}{-5} = -\sqrt{3}$　　$\cot \theta = \dfrac{x}{y} = \dfrac{-5}{5\sqrt{2}} = -\dfrac{\sqrt{2}}{2}$

Exercises

The terminal side of θ in standard position contains each point. Find the exact values of the six trigonometric functions of θ.

1. $(8, 4)$

2. $(4, 4)$

3. $(0, 4)$

4. $(6, 2)$

13-3　Study Guide and Intervention　(continued)

Trigonometric Functions of General Angles

Trigonometric Functions with Reference Angles If θ is a nonquadrantal angle in standard position, its **reference angle** θ' is defined as the acute angle formed by the terminal side of θ and the x-axis.

Reference Angle Rule	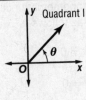 $\theta' = \theta$	$\theta' = 180° - \theta$ $(\theta' = \pi - \theta)$	Quadrant III $\theta' = \theta - 180°$ $(\theta' = \theta - \pi)$	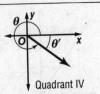 Quadrant IV $\theta' = 360° - \theta$ $(\theta' = 2\pi - \theta)$

Example 1　Sketch an angle of measure 205°. Then find its reference angle.

Because the terminal side of 205° lies in Quadrant III, the reference angle θ' is 205° − 180° or 25°.

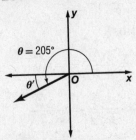

Example 2　Use a reference angle to find the exact value of $\cos \frac{3\pi}{4}$.

Because the terminal side of $\frac{3\pi}{4}$ lies in Quadrant II, the reference angle θ' is
$\pi - \frac{3\pi}{4}$ or $\frac{\pi}{4}$.
The cosine function is negative in Quadrant II.

$\cos \frac{3\pi}{4} = -\cos \frac{\pi}{4} = -\frac{\sqrt{2}}{2}$.

Exercises

Sketch each angle. Then find its reference angle.

1. 155°　　　　　**2.** 230°　　　　　**3.** $\frac{4\pi}{3}$　　　　　**4.** $-\frac{\pi}{6}$

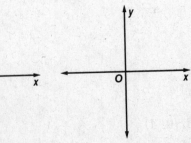

Find the exact value of each trigonometric function.

5. $\tan 330°$　　　**6.** $\cos \frac{11\pi}{4}$　　　**7.** $\cot 30°$　　　**8.** $\csc \frac{\pi}{4}$

13-4 Study Guide and Intervention

Law of Sines

Find the Area of a Triangle The area of any triangle is one half the product of the lengths of two sides and the sine of the included angle.

Area of a Triangle	area $= \frac{1}{2}bc \sin A$ area $= \frac{1}{2}ac \sin B$ area $= \frac{1}{2}ab \sin C$	

Example **Find the area of $\triangle ABC$ to the nearest tenth.**

In $\triangle ABC$, $a = 10$, $b = 14$, and $C = 40°$.

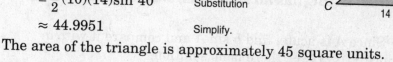

Area $= \frac{1}{2}ab \sin C$ Area formula

$= \frac{1}{2}(10)(14)\sin 40°$ Substitution

≈ 44.9951 Simplify.

The area of the triangle is approximately 45 square units.

Exercises

Find the area of $\triangle ABC$ to the nearest tenth, if necessary.

1.
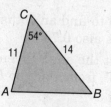

2.
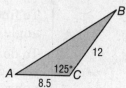

3.
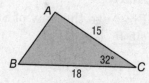

4.
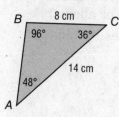

5.
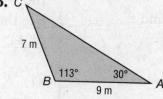

6.
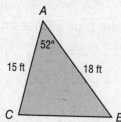

7. $A = 20°$, $c = 4$ cm, $b = 7$ cm

8. $C = 55°$, $a = 10$ m, $b = 15$ m

9. $B = 42°$, $c = 9$ ft, $a = 3$ ft

10. $c = 15$ in., $b = 13$ in., $A = 53°$

11. $a = 12$ cm, $b = 8$ cm, $C = 85°$

13-4 Study Guide and Intervention *(continued)*

Law of Sines

Use the Law of Sines to Solve Triangles You can use the Law of Sines to solve any triangle if you know the measures of two angles and any side opposite one of the angles, or the measures of two sides and the angle opposite one of them.

Law of Sines	$\dfrac{\sin A}{a} = \dfrac{\sin B}{b} = \dfrac{\sin C}{c}$

Possible Triangles Given Two Sides and One Opposite Angle	Suppose you are given a, b, and A for a triangle. If a is acute: $a < b \sin A$ $\Rightarrow$ no solution $a = b \sin A$ $\Rightarrow$ one solution $b > a > b \sin A$ $\Rightarrow$ two solutions $a > b$ $\Rightarrow$ one solution	If A is right or obtuse: $a \le b$ $\Rightarrow$ no solution $a > b$ $\Rightarrow$ one solution

Example Determine whether $\triangle ABC$ has *no* solutions, *one* solution, or *two* solutions. Then solve $\triangle ABC$.

a. $A = 48°$, $a = 11$, and $b = 16$ Since A is acute, find $b \sin A$ and compare it with a.
$b \sin A = 16 \sin 48° \approx 11.89$ Since $11 < 11.89$, there is no solution.

b. $A = 34°$, $a = 6$, $b = 8$
Since A is acute, find $b \sin A$ and compare it with a; $b \sin A = 8 \sin 34° \approx 4.47$. Since $8 > 6 > 4.47$, there are two solutions. Thus there are two possible triangles to solve.

Acute B	**Obtuse B**
First use the Law of Sines to find B. $\dfrac{\sin B}{8} = \dfrac{\sin 34°}{6}$ $\sin B \approx 0.7456$ $\quad B \approx 48°$	To find B you need to find an obtuse angle whose sine is also 0.7456. To do this, subtract the angle given by your calculator, 48°, from 180°. So B is approximately 132°.
The measure of angle C is about $180° - (34° + 48°)$ or about 98°.	The measure of angle C is about $180° - (34° + 132°)$ or about 14°.
Use the Law of Sines again to find c. $\dfrac{\sin 98°}{c} \approx \dfrac{\sin 34°}{6}$ $c \approx \dfrac{6 \sin 98°}{\sin 34°}$ $c \approx 10.6$	Use the Law of Sines to find c. $\dfrac{\sin 14°}{c} \approx \dfrac{\sin 34°}{6}$ $c \approx \dfrac{6 \sin 14°}{\sin 34°}$ $c \approx 2.6$

Exercises

Determine whether each triangle has *no* solution, *one* solution, or *two* solutions. Then solve the triangle. Round side lengths to the nearest tenth and angle measures to the nearest degree.

1. $A = 50°$, $a = 34$, $b = 40$ **2.** $A = 24°$, $a = 3$, $b = 8$ **3.** $A = 125°$, $a = 22$, $b = 15$

13-5 Study Guide and Intervention

Law of Cosines

Use Law of Cosines to Solve Triangles

Law of Cosines	Let $\triangle ABC$ be any triangle with a, b, and c representing the measures of the sides, and opposite angles with measures A, B, and C, respectively. Then the following equations are true.
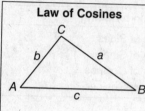	$a^2 = b^2 + c^2 - 2bc \cos A$ $b^2 = a^2 + c^2 - 2ac \cos B$ $c^2 = a^2 + b^2 - 2ab \cos C$

You can use the Law of Cosines to solve any triangle if you know the measures of two sides and the included angle (SAS case), or the measures of three sides (SSS case).

Example **Solve $\triangle ABC$.**

You are given the measures of two sides and the included angle.
Begin by using the Law of Cosines to determine c.

$c^2 = a^2 + b^2 - 2ab \cos C$

$c^2 = 28^2 + 15^2 - 2(28)(15)\cos 82°$

$c^2 \approx 892.09$

$c \approx 29.9$

Next you can use the Law of Sines to find the measure of angle A.

$\dfrac{\sin A}{a} = \dfrac{\sin C}{c}$

$\dfrac{\sin A}{28} \approx \dfrac{\sin 82°}{29.9}$

$\sin A \approx 0.9273$

$A \approx 68°$

The measure of B is about $180° - (82° + 68°)$ or about $30°$.

Exercises

Solve each triangle. Round side lengths to the nearest tenth and angle measures to the nearest degree.

1. $a = 14$, $c = 20$, $B = 38°$

2. $A = 60°$, $c = 17$, $b = 12$

3. $a = 4$, $b = 6$, $c = 3$

4. $A = 103°$, $b = 31$, $c = 52$

5. $a = 15$, $b = 26$, $C = 132°$

6. $a = 31$, $b = 52$, $c = 43$

13-5 Study Guide and Intervention *(continued)*

Law of Cosines

Choose a Method to Solve Triangles

	Given	Begin by Using
Solving an Oblique Triangle	two angles and any side	Law of Sines
	two sides and an angle opposite one of them	Law of Sines
	two sides and their included angle	Law of Cosines
	three sides	Law of Cosines

Example Determine whether $\triangle ABC$ should be solved by beginning with the Law of *Sines* or Law of *Cosines*. Then solve the triangle.

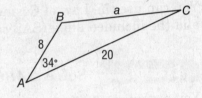

You are given the measures of two sides and their included angle, so use the Law of Cosines.

$a^2 = b^2 + c^2 - 2bc \cos A$ Law of Cosines

$a^2 = 20^2 + 8^2 - 2(20)(8) \cos 34°$ $b = 20, c = 8, A = 34°$

$a^2 \approx 198.71$ Use a calculator to simplify.

$a \approx 14.1$ Use a calculator to simplify.

Use the Law of Sines to find C.

$\dfrac{\sin C}{c} = \dfrac{\sin A}{a}$ Law of Sines

$\sin C \approx \dfrac{8 \sin 34°}{14.1}$ $c = 8, A = 34°, a \approx 14.1$

$C \approx 18°$ Use the $\sin^{-1}$ function.

The measure of angle B is approximately $180° - (34° + 18°)$ or about $128°$.

Exercises

Determine whether each triangle should be solved by beginning with the Law of *Sines* or Law of *Cosines*. Then solve the triangle.

1.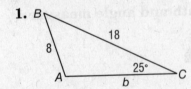

2.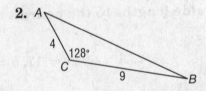

3.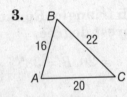

4. $A = 58°$, $a = 12$, $b = 8$ **5.** $a = 28$, $b = 35$, $c = 20$ **6.** $A = 82°$, $B = 44°$, $b = 11$

13-6 Study Guide and Intervention
Circular Functions

Circular Functions

Definition of Sine and Cosine	If the terminal side of an angle θ in standard position intersects the unit circle at $P(x, y)$, then $\cos \theta = x$ and $\sin \theta = y$. Therefore, the coordinates of P can be written as $P(\cos \theta, \sin \theta)$.

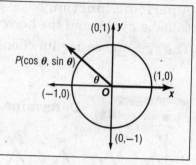

Example The terminal side of angle θ in standard position intersects the unit circle at $P\left(-\frac{5}{6}, \frac{\sqrt{11}}{6}\right)$. Find $\cos \theta$ and $\sin \theta$.

$P\left(-\frac{5}{6}, \frac{\sqrt{11}}{6}\right) = P(\cos \theta, \sin \theta)$, so $\cos \theta = -\frac{5}{6}$ and $\sin \theta = \frac{\sqrt{11}}{6}$.

Exercises

The terminal side of angle θ in standard position intersects the unit circle at each point P. Find $\cos \theta$ and $\sin \theta$.

1. $P\left(-\frac{\sqrt{3}}{2}, \frac{1}{2}\right)$

2. $P(0, -1)$

3. $P\left(-\frac{2}{3}, \frac{\sqrt{5}}{3}\right)$

4. $P\left(-\frac{4}{5}, -\frac{3}{5}\right)$

5. $P\left(\frac{1}{6}, -\frac{\sqrt{35}}{6}\right)$

6. $P\left(\frac{\sqrt{7}}{4}, \frac{3}{4}\right)$

7. P is on the terminal side of $\theta = 45°$.

8. P is on the terminal side of $\theta = 120°$.

9. P is on the terminal side of $\theta = 240°$.

10. P is on the terminal side of $\theta = 330°$.

13-6 Study Guide and Intervention (continued)

Circular Functions

Periodic Functions

A **periodic function** has *y*-values that repeat at regular intervals. One complete pattern is called a **cycle**, and the horizontal length of one cycle is called a **period**.

The sine and cosine functions are periodic; each has a period of 360° or 2π radians.

Example 1 Determine the period of the function.

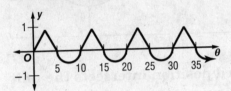

The pattern of the function repeats every 10 units, so the period of the function is 10.

Example 2 Find the exact value of each function.

a. sin 855°

$$\sin 855° = \sin (135° + 720°)$$
$$= \sin 135° \text{ or } \frac{\sqrt{2}}{2}$$

b. $\cos \left(\frac{31\pi}{6}\right)$

$$\cos \left(\frac{31\pi}{6}\right) = \cos \left(\frac{7\pi}{6} + 4\pi\right)$$
$$= \cos \frac{7\pi}{6} \text{ or } -\frac{\sqrt{3}}{2}$$

Exercises

Determine the period of each function.

1.

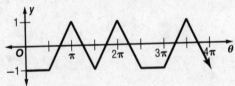

2.

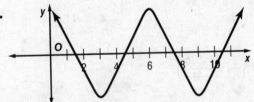

Find the exact value of each function.

3. sin (−510°)

4. sin 495°

5. $\cos \left(-\frac{5\pi}{2}\right)$

6. $\sin \left(\frac{5\pi}{3}\right)$

7. $\cos \left(\frac{11\pi}{4}\right)$

8. $\sin \left(-\frac{3\pi}{4}\right)$

13-7 Study Guide and Intervention

Graphing Trigonometric Functions

Sine, Cosine, and Tangent Functions Trigonometric functions can be graphed on the coordinate plane. Graphs of periodic functions have repeating patterns, or *cycles*; the horizontal length of each cycle is the *period*. The **amplitude** of the graph of a sine or cosine function equals half the difference between the maximum and minimum values of the function. Tangent is a trigonometric function that has asymptotes when graphed.

Sine, Cosine, and Tangent Functions	Parent Function	$y = \sin \theta$	$y = \cos \theta$	$y = \tan \theta$
	Domain	{all real numbers}	{all real numbers}	$\{\theta \mid \theta \neq 90 + 180n, n \text{ is an integer}\}$
	Range	$\{y \mid -1 \leq y \leq 1\}$	$\{y \mid -1 \leq y \leq 1\}$	{all real numbers}
	Amplitude	1	1	undefined
	Period	360°	360°	180°

Example Find the amplitude and period of each function. Then graph the function.

a. $y = 4 \cos \dfrac{\theta}{3}$

First, find the amplitude.

$|a| = |4|$, so the amplitude is 4.

Next find the period.

$\dfrac{360°}{\left|\frac{1}{3}\right|} = 1080°$

Use the amplitude and period to help graph the function.

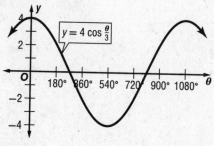

b. $y = \dfrac{1}{2} \tan 2\theta$

The amplitude is not defined, and the period is 90°.

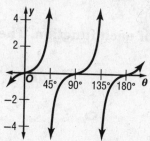

Exercises

Find the amplitude, if it exists, and period of each function. Then graph the function.

1. $y = -4 \sin \theta$

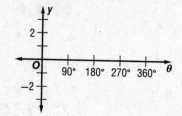

2. $y = 2 \tan \dfrac{\theta}{2}$

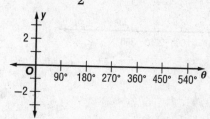

13-7 Study Guide and Intervention *(continued)*

Graphing Trigonometric Functions

Graphs of Other Trigonometric Functions The graphs of the cosecant, secant, and cotangent functions are related to the graphs of the sine, cosine, and tangent functions.

	Parent Function	$y = \csc \theta$	$y = \sec \theta$	$y = \cot \theta$
Cosecant, Secant, and Cotangent Functions	Domain	$\{\theta \mid \theta \neq 180n, n$ is an integer$\}$	$\{\theta \mid \theta \neq 90 + 180n, n$ is an integer$\}$	$\{\theta \mid \theta \neq 180n, n$ is an integer$\}$
	Range	$\{y \mid -1 > y$ or $y > 1\}$	$\{y \mid -1 > y$ or $y > 1\}$	$\{$all real numbers$\}$
	Amplitude	undefined	undefined	undefined
	Period	360°	360°	180°

Example   **Find the period of $y = \dfrac{1}{2}\csc \theta$. Then graph the function.**

Since $\dfrac{1}{2}\csc \theta$ is a reciprocal of $\dfrac{1}{2}\sin \theta$, the graphs
have the same period, 360°. The vertical
asymptotes occur at the points where $\dfrac{1}{2}\sin \theta = 0$.
So, the asymptotes are at $\theta = 0°$, $\theta = 180°$,
and $\theta = 360°$. Sketch $y = \dfrac{1}{2}\sin \theta$ and use it to graph
$y = \dfrac{1}{2}\csc \theta$.

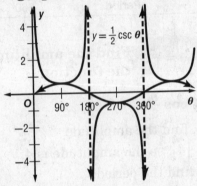

Exercises

Find the period of each function. Then graph the function.

1. $y = \cot 2\theta$

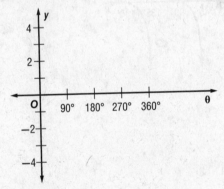

2. $y = \sec 3\theta$

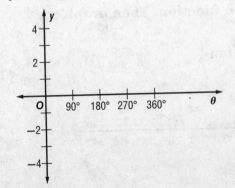

180

13-8 Study Guide and Intervention

Translations of Trigonometric Graphs

Horizontal Translations When a constant is subtracted from the angle measure in a trigonometric function, a **phase shift** of the graph results.

Phase Shift	The phase shift of the graphs of the functions $y = a \sin b(\theta - h)$, $y = a \cos b(\theta - h)$, and $y = a \tan b(\theta - h)$ is h, where $b > 0$. If $h > 0$, the shift is h units to the right. If $h < 0$, the shift is h units to the left.

Example State the amplitude, period, and phase shift for $y = \frac{1}{2} \cos 3\left(\theta - \frac{\pi}{2}\right)$. Then graph the function.

Amplitude: $|a| = \left|\frac{1}{2}\right|$ or $\frac{1}{2}$

Period: $\frac{2\pi}{|b|} = \frac{2\pi}{|3|}$ or $\frac{2\pi}{3}$

Phase Shift: $h = \frac{\pi}{2}$

The phase shift is to the right since $\frac{\pi}{2} > 0$.

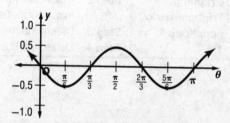

Exercises

State the amplitude, period, and phase shift for each function. Then graph the function.

1. $y = 2 \sin (\theta + 60°)$

2. $y = \tan \left(\theta - \frac{\pi}{2}\right)$

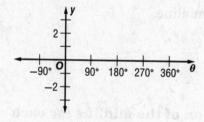

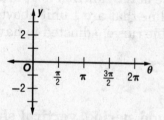

3. $y = 3 \cos (\theta - 45°)$

4. $y = \frac{1}{2} \sin 3\left(\theta - \frac{\pi}{3}\right)$

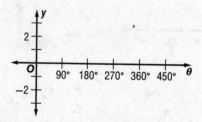

13-8 Study Guide and Intervention (continued)

Translations of Trigonometric Graphs

Vertical Translations When a constant is added to a trigonometric function, the graph is shifted vertically.

Vertical Shift	The vertical shift of the graphs of the functions $y = a \sin b(\theta - h) + k$, $y = a \cos b(\theta - h) + k$, and $y = a \tan b(\theta - h) + k$ is k. If $k > 0$, the shift is k units up. If $k < 0$, the shift is k units down.

The **midline** of a vertical shift is $y = k$.

Graphing Trigonometric Functions	**Step 1** Determine the vertical shift, and graph the midline. **Step 2** Determine the amplitude, if it exists. Use dashed lines to indicate the maximum and minimum values of the function. **Step 3** Determine the period of the function and graph the appropriate function. **Step 4** Determine the phase shift and translate the graph accordingly.

Example State the amplitude, period, vertical shift, and equation of the midline for $y = \cos 2\theta - 3$. Then graph the function.

Amplitude: $|a| = |1|$ or 1

Period: $\dfrac{2\pi}{|b|} = \dfrac{2\pi}{|2|}$ or π

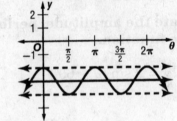

Vertical Shift: $k = -3$, so the vertical shift is 3 units down.

The equation of the midline is $y = -3$.

Since the amplitude of the function is 1, draw dashed lines parallel to the midline that are 1 unit above and below the midline. Then draw the cosine curve, adjusted to have a period of π.

Exercises

State the amplitude, period, vertical shift, and equation of the midline for each function. Then graph the function.

1. $y = \dfrac{1}{2} \cos \theta + 2$

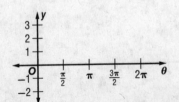

2. $y = 3 \sin \theta - 2$

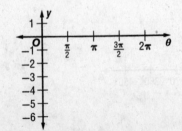

13-9 Study Guide and Intervention
Inverse Trigonometric Functions

Inverse Trigonometric Functions If you know the value of a trigonometric function for an angle, you can use the *inverse* to find the angle. If you restrict the function's domain, then the inverse is a function. The values in this restricted domain are called **principal values**.

Principal Values of Sine, Cosine, and Tangent	$y = \text{Sin } x$ if and only if $y = \sin x$ and $-\frac{\pi}{2} \le x \le \frac{\pi}{2}$. $y = \text{Cos } x$ if and only if $y = \cos x$ and $0 \le x \le \pi$. $y = \text{Tan } x$ if and only if $y = \tan x$ and $-\frac{\pi}{2} \le x \le \frac{\pi}{2}$.
Inverse Sine, Cosine, and Tangent	Given $y = \text{Sin } x$, the inverse sine function is defined by $y = \text{Sin}^{-1} x$ or $y = \text{Arcsin } x$. Given $y = \text{Cos } x$, the inverse cosine function is defined by $y = \text{Cos}^{-1} x$ or $y = \text{Arccos } x$. Given $y = \text{Tan } x$, the inverse tangent function is given by $y = \text{Tan}^{-1} x$ or $y = \text{Arctan } x$.

Example 1 **Find the value of $\text{Sin}^{-1}\left(\frac{\sqrt{3}}{2}\right)$. Write angle measures in degrees and radians.**

Find the angle θ for $-\frac{\pi}{2} \le \theta \le \frac{\pi}{2}$ that has a sine value of $\frac{\sqrt{3}}{2}$.

Using a unit circle, the point on the circle that has y-coordinate of $\frac{\sqrt{3}}{2}$ is $\frac{\pi}{3}$ or $60°$.

So, $\text{Sin}^{-1}\left(\frac{\sqrt{3}}{2}\right) = \frac{\pi}{3}$ or $60°$.

Example 2 **Find $\tan\left(\text{Sin}^{-1}\frac{1}{2}\right)$. Round to the nearest hundredth.**

Let $\theta = \text{Sin}^{-1}\frac{1}{2}$. Then $\text{Sin } \theta = \frac{1}{2}$ with $-\frac{\pi}{2} < \theta < \frac{\pi}{2}$. The value $\theta = \frac{\pi}{6}$ satisfies both conditions. $\tan\frac{\pi}{6} = \frac{\sqrt{3}}{3}$ so $\tan\left(\text{Sin}^{-1}\frac{1}{2}\right) = \frac{\sqrt{3}}{3}$.

Exercises

Find each value. Write angle measures in degrees and radians.

1. $\text{Cos}^{-1}\left(\frac{\sqrt{3}}{2}\right)$

2. $\text{Sin}^{-1}\left(-\frac{\sqrt{3}}{2}\right)$

3. $\text{Arccos}\left(-\frac{1}{2}\right)$

4. $\text{Arctan } \sqrt{3}$

5. $\text{Arccos}\left(-\frac{\sqrt{2}}{2}\right)$

6. $\text{Tan}^{-1}(-1)$

Find each value. Round to the nearest hundredth if necessary.

7. $\cos\left[\text{Sin}^{-1}\left(-\frac{\sqrt{2}}{2}\right)\right]$

8. $\tan\left[\text{Arcsin}\left(-\frac{5}{7}\right)\right]$

9. $\sin\left(\text{Tan}^{-1}\frac{5}{12}\right)$

10. $\text{Cos }[\text{Arcsin }(-0.7)]$

11. $\cos(\text{Arctan } 5)$

12. $\sin(\text{Cos}^{-1} 0.3)$

13-9 Study Guide and Intervention *(continued)*

Inverse Trigonometric Functions

Solve Equations by Using Inverses You can rewrite trigonometric equations to solve for the measure of an angle.

> **Example** **Solve the equation Sin $\theta = -0.25$. Round to the nearest tenth if necessary.**
>
> The sine of angle θ is -0.25. This can be written as Arcsin$(-0.25) = \theta$.
>
> Use a calculator to solve.
>
> **KEYSTROKES:** 2nd SIN⁻¹ (−) .25 ENTER −14.47751219
>
> So, $\theta \approx -14.5°$

Exercises

Solve each equation. Round to the nearest tenth if necessary.

1. Sin $\theta = 0.8$

2. Tan $\theta = 4.5$

3. Cos $\theta = 0.5$

4. Cos $\theta = -0.95$

5. Sin $\theta = -0.1$

6. Tan $\theta = -1$

7. Cos $\theta = 0.52$

8. Cos $\theta = -0.2$

9. Sin $\theta = 0.35$

10. Tan $\theta = 8$

14-1 Study Guide and Intervention

Trigonometric Identities

Find Trigonometric Values A **trigonometric identity** is an equation involving trigonometric functions that is true for all values for which every expression in the equation is defined.

Basic Trigonometric Identities	Quotient Identities	$\tan \theta = \dfrac{\sin \theta}{\cos \theta}$ $\cot \theta = \dfrac{\cos \theta}{\sin \theta}$	
	Reciprocal Identities	$\csc \theta = \dfrac{1}{\sin \theta}$ $\sec \theta = \dfrac{1}{\cos \theta}$ $\cot \theta = \dfrac{1}{\tan \theta}$	
	Pythagorean Identities	$\cos^2 \theta + \sin^2 \theta = 1$ $\tan^2 \theta + 1 = \sec^2 \theta$ $\cot^2 \theta + 1 = \csc^2 \theta$	

Example Find the exact value of $\cot \theta$ if $\csc \theta = -\dfrac{11}{5}$ and $180° < \theta < 270°$.

$\cot^2 \theta + 1 = \csc^2 \theta$ — Trigonometric identity

$\cot^2 \theta + 1 = \left(-\dfrac{11}{5}\right)^2$ — Substitute $-\dfrac{11}{5}$ for $\csc \theta$.

$\cot^2 \theta + 1 = \dfrac{121}{25}$ — Square $-\dfrac{11}{5}$.

$\cot^2 \theta = \dfrac{96}{25}$ — Subtract 1 from each side.

$\cot \theta = \pm\dfrac{4\sqrt{6}}{5}$ — Take the square root of each side.

Since θ is in the third quadrant, $\cot \theta$ is positive. Thus, $\cot \theta = \dfrac{4\sqrt{6}}{5}$.

Exercises

Find the exact value of each expression if $0° < \theta < 90°$.

1. If $\cot \theta = 4$, find $\tan \theta$.

2. If $\cos \theta = \dfrac{\sqrt{3}}{2}$, find $\csc \theta$.

3. If $\sin \theta = \dfrac{3}{5}$, find $\cos \theta$.

4. If $\sin \theta = \dfrac{1}{3}$, find $\sec \theta$.

5. If $\tan \theta = \dfrac{4}{3}$, find $\cos \theta$.

6. If $\sin \theta = \dfrac{3}{7}$, find $\tan \theta$.

Find the exact value of each expression if $90° < \theta < 180°$.

7. If $\cos \theta = -\dfrac{7}{8}$, find $\sec \theta$.

8. If $\csc \theta = \dfrac{12}{5}$, find $\cot \theta$.

Find the exact value of each expression if $270° < \theta < 360°$.

9. If $\cos \theta = \dfrac{6}{7}$, find $\sin \theta$.

10. If $\csc \theta = -\dfrac{9}{4}$, find $\sin \theta$.

14-1 Study Guide and Intervention (continued)

Trigonometric Identities

Simplify Expressions The simplified form of a trigonometric expression is written as a numerical value or in terms of a single trigonometric function, if possible. Any of the trigonometric identities can be used to simplify expressions containing trigonometric functions.

Example 1 Simplify $(1 - \cos^2 \theta) \sec \theta \cot \theta + \tan \theta \sec \theta \cos^2 \theta$.

$$(1 - \cos^2 \theta) \sec \theta \cot \theta + \tan \theta \sec \theta \cos^2 \theta = \sin^2 \theta \cdot \frac{1}{\cos \theta} \cdot \frac{\cos \theta}{\sin \theta} + \frac{\sin \theta}{\cos \theta} \cdot \frac{1}{\cos \theta} \cdot \cos^2 \theta$$

$$= \sin \theta + \sin \theta$$

$$= 2 \sin \theta$$

Example 2 Simplify $\dfrac{\sec \theta \cdot \cot \theta}{1 - \sin \theta} - \dfrac{\csc \theta}{1 + \sin \theta}$.

$$\frac{\sec \theta \cdot \cot \theta}{1 - \sin \theta} - \frac{\csc \theta}{1 + \sin \theta} = \frac{\frac{1}{\cos \theta} \cdot \frac{\cos \theta}{\sin \theta}}{1 + \sin \theta} - \frac{\frac{1}{\sin \theta}}{1 + \sin \theta}$$

$$= \frac{\frac{1}{\sin \theta}(1 + \sin \theta) - \frac{1}{\sin \theta}(1 - \sin \theta)}{(1 - \sin \theta)(1 + \sin \theta)}$$

$$= \frac{\frac{1}{\sin \theta} + 1 - \frac{1}{\sin \theta} + 1}{1 - \sin^2 \theta}$$

$$= \frac{2}{\cos^2 \theta} \text{ or } 2 \sec^2 \theta$$

Exercises

Simplify each expression.

1. $\dfrac{\tan \theta \cdot \csc \theta}{\sec \theta}$

2. $\dfrac{\sin \theta \cdot \cot \theta}{\sec^2 \theta - \tan^2 \theta}$

3. $\dfrac{\sin^2 \theta - \cot \theta \cdot \tan \theta}{\cot \theta \cdot \sin \theta}$

4. $\dfrac{\cos \theta}{\sec \theta - \tan \theta}$

5. $\dfrac{\tan \theta \cdot \cos \theta}{\sin \theta} + \cot \theta \cdot \sin \theta \cdot \tan \theta \cdot \csc \theta$

6. $\dfrac{\csc^2 \theta - \cot^2 \theta}{\tan \theta \cdot \cos \theta}$

7. $3 \tan \theta \cdot \cot \theta + 4 \sin \theta \cdot \csc \theta + 2 \cos \theta \cdot \sec \theta$

8. $\dfrac{1 - \cos^2 \theta}{\tan \theta \cdot \sin \theta}$

14-2 Study Guide and Intervention

Verifying Trigonometric Identities

Transform One Side of an Equation Use the basic trigonometric identities along with the definitions of the trigonometric functions to verify trigonometric identities. Often it is easier to begin with the more complicated side of the equation and transform that expression into the form of the simpler side.

Example | **Verify that each equation is an identity.**

a. $\dfrac{\sin \theta}{\cot \theta} - \sec \theta = -\cos \theta$

Transform the left side.

$\dfrac{\sin \theta}{\cot \theta} - \sec \theta \overset{?}{=} -\cos \theta$

$\dfrac{\sin \theta}{\frac{\cos \theta}{\sin \theta}} - \dfrac{1}{\cos \theta} \overset{?}{=} -\cos \theta$

$\dfrac{\sin^2 \theta}{\cos \theta} - \dfrac{1}{\cos \theta} \overset{?}{=} -\cos \theta$

$\dfrac{\sin^2 - 1}{\cos \theta} \overset{?}{=} -\cos \theta$

$\dfrac{-\cos^2 \theta}{\cos \theta} \overset{?}{=} -\cos \theta$

$-\cos \theta = -\cos \theta \checkmark$

b. $\dfrac{\tan \theta}{\csc \theta} + \cos \theta = \sec \theta$

Transform the left side.

$\dfrac{\tan \theta}{\csc \theta} + \cos \theta \overset{?}{=} \sec \theta$

$\dfrac{\frac{\sin \theta}{\cos \theta}}{\frac{1}{\sin \theta}} + \cos \theta \overset{?}{=} \sec \theta$

$\dfrac{\sin^2 \theta}{\cos \theta} + \cos \theta \overset{?}{=} \sec \theta$

$\dfrac{\sin^2 \theta + \cos^2 \theta}{\cos \theta} \overset{?}{=} \sec \theta$

$\dfrac{1}{\cos \theta} \overset{?}{=} \sec \theta$

$\sec \theta = \sec \theta \checkmark$

Exercises

Verify that each equation is an identity.

1. $1 + \csc^2 \theta \cdot \cos^2 \theta = \csc^2 \theta$

2. $\dfrac{\sin \theta}{1 - \cos \theta} - \dfrac{\cot \theta}{1 + \cos \theta} = \dfrac{1 - \cos^3 \theta}{\sin^3 \theta}$

14-2 Study Guide and Intervention *(continued)*

Verifying Trigonometric Identities

Transform Each Side of an Equation The following techniques can be helpful in verifying trigonometric identities.

- Substitute one or more basic identities to simplify an expression.
- Factor or multiply to simplify an expression.
- Multiply both numerator and denominator by the same trigonometric expression.
- Write each side of the identity in terms of sine and cosine only. Then simplify each side.

Example Verify that $\dfrac{\tan^2 \theta + 1}{\sin \theta \cdot \tan \theta \cdot \sec \theta + 1} = \sec^2 \theta - \tan^2 \theta$ is an identity.

$$\dfrac{\tan^2 \theta + 1}{\sin \theta \cdot \tan \theta \cdot \sec \theta + 1} \overset{?}{=} \sec^2 \theta - \tan^2 \theta$$

$$\dfrac{\sec^2 \theta}{\sin \theta \cdot \dfrac{\sin \theta}{\cos \theta} \cdot \dfrac{1}{\cos \theta} + 1} \overset{?}{=} \dfrac{1}{\cos^2 \theta} - \dfrac{\sin^2 \theta}{\cos^2 \theta}$$

$$\dfrac{\dfrac{1}{\cos^2 \theta}}{\dfrac{\sin^2 \theta}{\cos^2 \theta} + 1} \overset{?}{=} \dfrac{1 - \sin^2 \theta}{\cos^2 \theta}$$

$$\dfrac{\dfrac{1}{\cos^2 \theta}}{\dfrac{\sin^2 \theta + \cos^2 \theta}{\cos^2 \theta}} \overset{?}{=} \dfrac{\cos^2 \theta}{\cos^2 \theta}$$

$$\dfrac{1}{\sin^2 \theta + \cos^2 \theta} \overset{?}{=} 1$$

$$1 = 1 \checkmark$$

Exercises

Verify that each equation is an identity.

1. $\csc \theta \cdot \sec \theta = \cot \theta + \tan \theta$

2. $\dfrac{\tan^2 \theta}{1 - \cos^2 \theta} = \dfrac{\sec \theta}{\cos \theta}$

3. $\dfrac{\cos \theta \cdot \cot \theta}{\sin \theta} = \dfrac{\csc \theta}{\sin \theta \cdot \sec^2 \theta}$

4. $\dfrac{\csc^2 \theta - \cot^2 \theta}{\sec^2 \theta} = \cot^2 \theta(1 - \cos^2 \theta)$

14-3 Study Guide and Intervention

Sum and Difference of Angles Identities

Sum and Difference Identities The following formulas are useful for evaluating an expression like sin 15° from the known values of sine and cosine of 60° and 45°.

Sum and Difference of Angles	The following identities hold true for all values of α and β. $\cos (\alpha \pm \beta) = \cos \alpha \cdot \cos \beta \mp \sin \alpha \cdot \sin \beta$ $\sin (\alpha \pm \beta) = \sin \alpha \cdot \cos \beta \pm \cos \alpha \cdot \sin \beta$

Example Find the exact value of each expression.

a. cos 345°

$\cos 345° = \cos (300° + 45°)$

$= \cos 300° \cdot \cos 45° - \sin 300° \cdot \sin 45°$

$= \dfrac{1}{2} \cdot \dfrac{\sqrt{2}}{2} - \left(-\dfrac{\sqrt{3}}{2}\right) \cdot \dfrac{\sqrt{2}}{2}$

$= \dfrac{\sqrt{2} + \sqrt{6}}{4}$

b. sin (−105°)

$\sin (-105°) = \sin (45° - 150°)$

$= \sin 45° \cdot \cos 150° - \cos 45° \cdot \sin 150°$

$= \dfrac{\sqrt{2}}{2} \cdot \left(-\dfrac{\sqrt{3}}{2}\right) - \dfrac{\sqrt{2}}{2} \cdot \dfrac{1}{2}$

$= -\dfrac{\sqrt{2} + \sqrt{6}}{4}$

Exercises

Find the exact value of each expression.

1. sin 105°

2. cos 285°

3. cos (−75°)

4. cos (−165°)

5. sin 195°

6. cos 420°

7. sin (−75°)

8. cos 135°

9. cos (−15°)

10. sin 345°

11. cos (−105°)

12. sin 495°

14-3 Study Guide and Intervention (continued)

Sum and Difference of Angles Identities

Verify Trigonometric Identities You can also use the sum and difference of angles formulas to verify identities.

Example 1 Verify that $\cos\left(\theta + \frac{3\pi}{2}\right) = \sin\theta$ is an identity.

$\cos\left(\theta + \frac{3\pi}{2}\right) \stackrel{?}{=} \sin\theta$ Original equation

$\cos\theta \cdot \cos\frac{3\pi}{2} - \sin\theta \cdot \sin\frac{3\pi}{2} \stackrel{?}{=} \sin\theta$ Sum of Angles Formula

$\cos\theta \cdot 0 - \sin\theta \cdot (-1) \stackrel{?}{=} \sin\theta$ Evaluate each expression.

$\sin\theta = \sin\theta \checkmark$ Simplify.

Example 2 Verify that $\sin\left(\theta - \frac{\pi}{2}\right) + \cos(\theta + \pi) = -2\cos\theta$ is an identity.

$\sin\left(\theta - \frac{\pi}{2}\right) + \cos(\theta + \pi) \stackrel{?}{=} -2\cos\theta$ Original equation

$\sin\theta \cdot \cos\frac{\pi}{2} - \cos\theta \cdot \sin\frac{\pi}{2} + \cos\theta \cdot \cos\pi - \sin\theta \cdot \sin\pi \stackrel{?}{=} -2\cos\theta$ Sum and Difference of Angles Formulas

$\sin\theta \cdot 0 - \cos\theta \cdot 1 + \cos\theta \cdot (-1) - \sin\theta \cdot 0 \stackrel{?}{=} -2\cos\theta$ Evaluate each expression.

$-2\cos\theta = -2\cos\theta \checkmark$ Simplify.

Exercises

Verify that each equation is an identity.

1. $\sin(90° + \theta) = \cos\theta$

2. $\cos(270° + \theta) = \sin\theta$

3. $\sin\left(\frac{2\pi}{3} - \theta\right) + \cos\left(\theta - \frac{5\pi}{6}\right) = \sin\theta$

4. $\cos\left(\frac{3\pi}{4} + \theta\right) - \left(\sin\theta - \frac{\pi}{4}\right) = -\sqrt{2}\sin\theta$

14-4 Study Guide and Intervention

Double-Angle and Half-Angle Identities

Double-Angle Identities

Double-Angle Identities	The following identities hold true for all values of θ. $\sin 2\theta = 2 \sin \theta \cdot \cos \theta$ $\qquad$ $\cos 2\theta = \cos^2 \theta - \sin^2 \theta$ $\cos 2\theta = 1 - 2 \sin^2 \theta$ $\cos 2\theta = 2 \cos^2 \theta - 1$

Example Find the exact values of $\sin 2\theta$, and $\cos 2\theta$ if $\sin \theta = -\dfrac{9}{10}$ and θ is between 180° and 270°.

First, find value of $\cos \theta$.

$\cos^2 \theta = 1 - \sin^2 \theta \qquad\qquad \cos^2 \theta + \sin^2 \theta = 1$

$\cos^2 \theta = 1 - \left(-\dfrac{9}{10}\right)^2 \qquad\quad \sin \theta = -\dfrac{9}{10}$

$\cos^2 \theta = \dfrac{19}{100}$

$\cos \theta = \pm\dfrac{\sqrt{19}}{10}$

Since θ is in the third quadrant, $\cos \theta$ is negative. Thus $\cos \theta = -\dfrac{\sqrt{19}}{10}$.

To find $\sin 2\theta$, use the identity $\sin 2\theta = 2 \sin \theta \cdot \cos \theta$.

$\sin 2\theta = 2 \sin \theta \cdot \cos \theta$.

$\quad = 2\left(-\dfrac{9}{10}\right)\left(-\dfrac{\sqrt{19}}{10}\right)$

$\quad = \dfrac{9\sqrt{19}}{50}$

The value of $\sin 2\theta$ is $\dfrac{9\sqrt{19}}{50}$.

To find $\cos 2\theta$, use the identity $\cos 2\theta = 1 - 2 \sin^2 \theta$.

$\cos 2\theta = 1 - 2 \sin^2 \theta$

$\quad = 1 - 2\left(-\dfrac{9}{10}\right)^2$

$\quad = -\dfrac{31}{50}$

The value of $\cos 2\theta$ is $= -\dfrac{31}{50}$.

Exercises

Find the exact values of $\sin 2\theta$ and $\cos 2\theta$.

1. $\sin \theta = \dfrac{1}{4}$, 0° < θ < 90°

2. $\sin \theta = -\dfrac{1}{8}$, 270° < θ < 360°

3. $\cos \theta = -\dfrac{3}{5}$, 180° < θ < 270°

4. $\cos \theta = -\dfrac{4}{5}$, 90° < θ < 180°

5. $\sin \theta = -\dfrac{3}{5}$, 270° < θ < 360°

6. $\cos \theta = -\dfrac{2}{3}$, 90° < θ < 180°

14-4 Study Guide and Intervention *(continued)*

Double-Angle and Half-Angle Identities

Half-Angle Identities

Half-Angle Identities	The following identities hold true for all values of a. $\sin \dfrac{\theta}{2} = \pm \sqrt{\dfrac{1 - \cos \theta}{2}} \qquad \cos \dfrac{\theta}{2} = \pm \sqrt{\dfrac{1 + \cos \theta}{2}}$

Example Find the exact value of $\sin \dfrac{\theta}{2}$ if $\sin \theta = \dfrac{2}{3}$ and θ is between 90° and 180°.

First find $\cos \theta$.

$\cos^2 \theta = 1 - \sin^2 \theta$ $\cos^2 \theta + \sin^2 \theta = 1$

$\cos^2 \theta = 1 - \left(\dfrac{2}{3}\right)^2$ $\sin \theta = \dfrac{2}{3}$

$\cos^2 \theta = \dfrac{5}{9}$ Simplify.

$\cos \theta = \pm \dfrac{\sqrt{5}}{3}$ Take the square root of each side.

Since θ is in the second quadrant, $\cos \theta = -\dfrac{\sqrt{5}}{3}$.

$\sin \dfrac{\theta}{2} = \pm \sqrt{\dfrac{1 - \cos \theta}{2}}$ Half-Angle formula

$= \pm \sqrt{\dfrac{1 - \left(-\dfrac{\sqrt{5}}{3}\right)}{2}}$ $\cos \theta = -\dfrac{\sqrt{5}}{3}$

$= \pm \sqrt{\dfrac{3 + \sqrt{5}}{6}}$ Simplify.

$= \pm \dfrac{\sqrt{18 + 6\sqrt{5}}}{6}$ Rationalize.

Since θ is between 90° and 180°, $\dfrac{\theta}{2}$ is between 45° and 90°. Thus $\sin \dfrac{\theta}{2}$ is positive and equals $\dfrac{\sqrt{18 + 6\sqrt{5}}}{6}$.

Exercises

Find the exact values of $\sin \dfrac{\theta}{2}$ and $\cos \dfrac{\theta}{2}$.

1. $\cos \theta = -\dfrac{3}{5}$, $180° < \theta < 270°$ **2.** $\cos \theta = -\dfrac{4}{5}$, $90° < \theta < 180°$

3. $\sin \theta = -\dfrac{3}{5}$, $270° < \theta < 360°$ **4.** $\cos \theta = -\dfrac{2}{3}$, $90° < \theta < 180°$

Find the exact value of each expression.

5. $\cos 22\dfrac{1}{2}°$ **6.** $\sin 67.5°$ **7.** $\cos \dfrac{7\pi}{8}$

14-5 Study Guide and Intervention
Solving Trigonometric Equations

Solve Trigonometric Equations You can use trigonometric identities to solve trigonometric equations, which are true for only certain values of the variable.

Example 1 Solve $4 \sin^2 \theta - 1 = 0$ if $0° < \theta < 360°$.

$4 \sin^2 \theta - 1 = 0$

$4 \sin^2 \theta = 1$

$\sin^2 \theta = \dfrac{1}{4}$

$\sin \theta = \pm \dfrac{1}{2}$

$\theta = 30°, 150°, 210°, 330°$

Example 2 Solve $\sin 2\theta + \cos \theta = 0$ for all values of θ. Give your answer in both radians and degrees.

$\sin 2\theta + \cos \theta = 0$

$2 \sin \theta \cos \theta + \cos \theta = 0$

$\cos \theta (2 \sin \theta + 1) = 0$

$\cos \theta = 0$ or $2 \sin \theta + 1 = 0$

$\sin \theta = -\dfrac{1}{2}$

$\theta = 90° + k \cdot 180°$; $\quad \theta = 210° + k \cdot 360°,$

$\theta = \dfrac{\pi}{2} + k \cdot \pi \quad\quad 330° + k \cdot 360°;$

$\theta = \dfrac{7\pi}{6} + k \cdot 2\pi,$

$\dfrac{11\pi}{6} + k \cdot 2\pi$

Exercises

Solve each equation for the given interval.

1. $2 \cos^2 \theta + \cos \theta = 1, 0 \le \theta < 2\pi$

2. $\sin^2 \theta \cos^2 \theta = 0, 0 \le \theta < 2\pi$

3. $\cos 2\theta = \dfrac{\sqrt{3}}{2}, 0° \le \theta < 360°$

4. $2 \sin \theta - \sqrt{3} = 0, 0 \le \theta < 2\pi$

Solve each equation for all values of θ if θ is measured in radians.

5. $4 \sin^2 \theta - 3 = 0$

6. $2 \cos \theta \sin \theta + \cos \theta = 0$

Solve each equation for all values of θ if θ is measured in degrees.

7. $\cos 2\theta + \sin^2 \theta = \dfrac{1}{2}$

8. $\tan 2\theta = -1$

14-5 Study Guide and Intervention *(continued)*

Solving Trigonometric Equations

Extraneous Solutions Some trigonometric equations have no solutions. For example, the equation $\sin \theta = 3$ has no solution because all values of $\sin \theta$ are between -1 and 1.

Example Solve $2 \cos^2 x + 3 \cos x - 2 = 0$ if $0 \le \theta \le 2\pi$.

$$2 \cos^2 \theta + \cos \theta - 1 = 0 \qquad \text{Original equation}$$

$$(\cos \theta + 2)(2 \cos \theta - 1) = 0 \qquad \text{Factor.}$$

$$\cos \theta + 2 = 0 \quad \text{or} \quad 2 \cos \theta - 1 = 0$$

$$\cos \theta = -2 \qquad 2 \cos \theta = 1$$

$$\cos \theta = \frac{1}{2}$$

$$\theta = \frac{\pi}{3}; \frac{5\pi}{3}$$

There is no solution to $\cos \theta = -2$ since all values of $\cos \theta$ are between -1 and 1, inclusive. The solutions are $\frac{\pi}{3}$ and $\frac{5\pi}{3}$.

Exercises

Solve each equation if $0 \le \theta \le 2\pi$.

1. $\sin^2 \theta + \dfrac{7}{2} \sin \theta + \dfrac{3}{2} = 0$

2. $2 \tan^4 \theta = \sec^2 \theta$

3. $8 \cos \theta = 4 \cos^2 \theta + 3$

4. $2 \csc^2 \theta = -(3 \csc \theta + 1)$

5. $2 \sin^2 \theta = 6 - 5\sqrt{2} \sin \theta$

6. $2 \cos^4 \theta + 9 \sin^2 \theta = 5$